SOLID POLYMER ELECTROLYTES

Fiona M. Gray

SOLID POLYMER ELECTROLYTES

*Fundamentals
and Technological
Applications*

Fiona M. Gray
Polymer Electrolyte Group Scotland
Department of Chemistry
University of St. Andrews
Scotland, UK

Library of Congress Cataloging-in-Publication Data

Gray, Fiona M.
    Solid polymer electrolytes : fundamentals and
technological applications / Fiona M. Gray.
        p. cm.
    Includes bibliographical references and index.
    ISBN 0-89573-772-8
    1. Polyelectrolytes. I. Title.
QD382.P64G73   1991
621.31'2424—dc20   91-13781
                                              CIP

British Library Cataloguing in Publication Data

Gray, Fiona M.
    Solid polymer electrolytes : fundamentals
    and technological applications.
    I. Title
    541.372

    ISBN 3-527-27925-3

© 1991 VCH Publishers, Inc.

Printed in the United States of America

ISBN 0-89573-772-8 VCH Publishers
ISBN 3-527-27925-3 VCH Verlagsgesellschaft

Printing History:
10 9 8 7 6 5 4 3 2 1

Published jointly by:

VCH Publishers, Inc.   VCH Verlagsgesellschaft mbH   VCH Publishers (UK) Ltd.
220 East 23rd Street   P.O. Box 10 11 61   8 Wellington Court
Suite 909   D-6940 Weinheim   Cambridge CB1 1HZ
New York, NY 10010   Federal Republic of Germany   United Kingdom

Preface

It has been known for many years that polyethers and various salts are capable of direct interaction and, indeed, these complexing properties have been widely used in organometallic chemistry. In 1973, Dr. P. V. Wright first reported the conducting properties of "solvent-free" poly(ethylene oxide)–salt systems, but it was in 1978 that Professor M. B. Armand highlighted the potential of these materials as a new class of solid electrolyte for energy storage applications.

With general concerns for future energy generation and conservation, a rapid growth in research and development relating to these new "polymer electrolytes" has ensued. The impetus for studying the fundamental aspects of polymer–salt systems has come about largely through the desire to develop thin-film rechargeable lithium batteries based on these materials. With diversification of research into areas other than lithium-based polymer electrolytes, it has become apparent that there is potential for exploitation in many other energy-related applications.

Over the past 3 to 4 years a number of general articles as well as more specialist reviews on specific areas relating to polymer electrolytes have been published. With interest spreading in both the academic and commercial fields, it seemed timely to prepare a book that would give a comprehensive account of the current state of knowledge in both the fundamental and applied aspects of these solid electrolytes.

Chapter 1 deals with the significant advances being made toward realizing practical devices. The major part of this covers lithium secondary battery research and development and scaleup projections of performance characteristics. Applications of polymer electrolytes in electrochromic displays and windows is an area that has recently received much attention and is also discussed. Aspects of polymer electrolyte formation and their phase behavior are described in Chapters 2 to 4. Chapter 5 deals with various aspects of the ionic conductivity. Chapters 6 and 7 concentrate on new materials; "designed" polymeric hosts for enhanced conductivity and non-alkali metal-based electrolytes. Chapters 8 and 9 concentrate on aspects relating to ion transport in polymer electrolytes. Theoretical models and experimental techniques relating to microscopic dynamic properties are described in Chapter 8. Chapter 9 focuses on the unresolved problems concerning ion–ion interactions, the nature of the charge carriers, and mode of charge transport. The significance and interpretation of experimental data in relation to these factors have also been highlighted. Finally, in Chapter 10, the electrolyte–lithium and electrolyte–intercalation cathode interfaces of polymer electrolyte-based cells are discussed.

It is hoped that this book will serve as a useful all-round introduction to those new to the field, as well as give the more familiar a comprehensive account and objective analysis of the subject.

I would like to take this opportunity to thank Professor Colin A. Vincent and other members of the Polymer Electrolyte Group Scotland (PEGS) for stimulating discussions which have been of great assistance in the preparation of this book.

Fiona M. Gray

Table of Contents

Chapter 7.   Further Developments in Polymer Electrolyte Materials

Chapter 8.   Transport Properties: Effects of Dynamic Disorder

CHAPTER 1

Polymer Electrolyte-Based Devices

1.1. Introduction

The production, storage, and distribution of energy are among the main concerns of modern industry and society. The development of new solid materials for both electrolyte and electrode applications is creating opportunities for new types of electrical power generation and storage systems which may themselves, in turn, revolutionize many industrial areas. The use of alternative energy sources that generate electricity on an intermittent basis, for example, solar energy and wind power, requires low-cost, high-efficiency electricity storage systems. Space development, creation of new types of memory, and new computer architecture, along with biomedical devices and microsensors for controlling atmospheric pollution, are all areas that could benefit from the development of solid-state ionic conductors. Although development of batteries for microelectronics is already very promising, the prospects of large-scale secondary power sources for electric vehicle traction are also encouraging.

Progress to date in solid-state ionics is largely the result of developments in two categories of materials: insertion compounds and fast ionic conductors. Polymer electrolytes represent the newest class of solid ionics. They contrast sharply with the usual solid ionic materials based on ceramics, glasses, or inorganic crystals with respect to the mode of charge transport (polymer electrolytes transport charge well only above their glass transition temperature) and the value of the ionic conductivity, which is of the order of 100 to 1000 times lower than for the latter materials. This drawback is compensated by a number of factors. For example, the polymer electrolytes can be formed into very thin films of large surface area giving high (>100 W dm^{-3}) power levels. Figure 1.1, which gives the relationship between conductivity or current density of the electrolyte and total thickness of the unit cell, shows how polymer electrolytes can in fact lead to volumetric power densities equal to those of liquid electrolytes or molten salts.[1] As the energy and power density requirements call for surfaces of the order of 15 to 25 m^2 kWh^{-1}, the flexible nature of these materials is essential, allowing space-efficient battery designs of variable dimensions to be constructed. From the point of view of electrochemical applications, polymer electrolyte flexibility is also important in that volume changes in the cell can be accommodated during cycling without physical degradation of the interfacial contacts, which is often observed in crystalline or vitreous solid electrolytes.

The essential feature that distinguishes a polymer electrolyte from low-molecular-weight solvent-based systems is that net ionic motion in polymer electrolytes takes

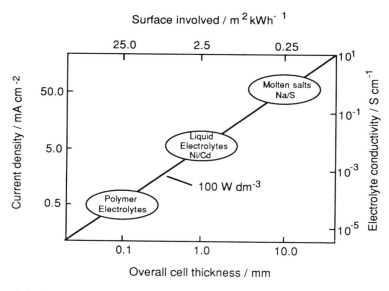

Figure 1.1. Characteristics of electrochemical cells for three battery technologies at power densities >100 W dm^{-3}. Reprinted from Ref. 1 by courtesy of Marcel Dekker, Inc.

place without long-range displacement of the solvent. Gel electrolytes (i.e., polymers containing a low-molecular-weight fraction) and polyelectrolytes again rely on an incorporated low-molecular-weight solvent medium to assist ionic transport; however, in a polymer electrolyte, no low-molecular-weight solvent is present and ion transport relies on local relaxation processes in the polymer chains which may provide liquidlike degrees of freedom, giving the polymer properties similar to those of a molecular liquid. The macroscopic properties that are similar to those of a solid are the result of chain entanglements and possibly crosslinking. Ion transport in polymer electrolytes is considered to take place by a combination of ion motion coupled to the local motion of polymer segments and inter- and intrapolymer transitions between ion coordinating sites. A simplified schematic representation of this is given in Figure 1.2. An important unresolved area in the field of polymer electrolytes concerns the role of the anion in ionic conductivity and the extent of ion association in these systems. It is generally accepted that anions are mobile, as, in some systems, it can be shown that net cation mobility is vanishingly small. As discussed in Chapter 9, however, it is likely that anions assist in cation transport. It may be envisaged that by forming, for example, ion pairs, triple ions, or higher aggregates, and with the assistance of polymer chain segmental motion, the ionic cluster may itself move or it may act as a transient center for the mobile species. This form of ion transport is depicted later in Figure 9.1.

To act as a successful polymer host, a polymer or the active part of a copolymer should generally have a minimum of three essential characteristics:

1. Atoms or groups of atoms with sufficient electron donor power to form coordinate bonds with cations

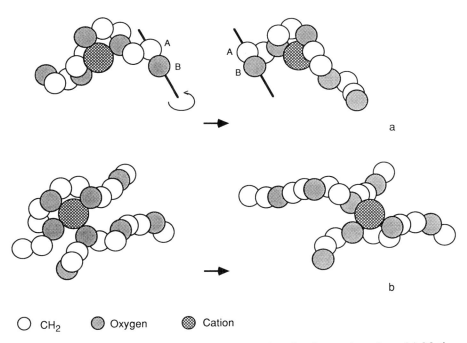

O CH₂   ◯ Oxygen   ⊗ Cation

Figure 1.2. Cation transport mechanism in a PEO-based polymer electrolyte. **(a)** Motion coupled to that of the polymer chain; lateral displacement brought about by 180° bond rotation at C—O bond AB. **(b)** First step in the transfer of a cation between chains. Anions may also be involved as part of either an ion pair or an ion triplet.

2. Low barriers to bond rotation so that segmental motion of the polymer chain can take place readily
3. A suitable distance between coordinating centers because the formation of multiple intrapolymer ion bonds appears to be important

Although a number of macromolecules satisfy these criteria, none have shown any advantage over poly(ethylene oxide) (PEO) which has, to date, been the most widely studied host polymer. The range of salts that may dissolve in PEO is large; however, attention has focused principally on a small group of lithium and sodium salts that form polymer electrolytes of potential commercial interest. As further opportunities for commercial exploitation of these materials have been realized, so the range of polymer–salt systems studied has increased, particularly in the areas of proton and multivalent cation conductors.

The future prospects for polymer electrolytes look promising because it has been appreciated that they form an ideal medium for a wide range of electrochemical processes. Other than primary and secondary batteries and ambient temperature fuel cells, practical applications for polymer electrolytes that are under consideration include

Electrochromic devices

Modified electrodes/sensors

Solid-state reference electrode systems

Supercapacitors

Thermoelectric generators

High-vacuum electrochemistry

Electrochemical switching

The primary driving force behind research into these materials has, however, been the development of secondary lithium batteries and therefore the major part of this chapter deals with developments in this field.

1.2. Major Advantages in Developing Polymer Electrolyte Batteries

Although a conductivity of the order of 10^{-4} S cm^{-1} may be obtained for a number of polymer electrolytes at various temperatures, this is not as high as may be desired. As was noted earlier, by configuring the electrolytes as very thin, large-area elements, the internal resistance of the resultant cell may be brought into an acceptable range. Also, by fabricating the anode and cathode structures as thin, high-area elements, the cells can be operated at relatively low values of current density, while still permitting the battery to operate at practical rates. Typical cell dimensions are ~15- to 30-μm-thick electrolyte, 25- to 50-μm-thick Li electrode, and 20- to 100-μm composite cathode. It is the combination of a unique battery cell structure with the properties of these solid electrolytes that can, in principle, permit high values of specific energy and power to be achieved.

The polymer electrolyte plays three important roles in the solid polymer electrolyte (SPE) battery.

1. It is a lithium ion carrier and, as noted earlier, can be formed into thin films to improve the energy density.
2. It acts as an electrode spacer, which eliminates the need to incorporate an inert porous separator.
3. It is a binder, which ensures good electrical contact with the electrodes. Thus there is no requirement for high-temperature operation to obtain liquid electrodes as, for example, in the Na–S battery. In addition, the flexibility and mechanical resilience puts SPEs at a major advantage over other all-solid-state cells based on crystalline or vitreous solid electrolytes in that electrode–electrolyte contact can be maintained at all times through charging and discharging.

Replacement of the liquid electrolyte by a plastic material means that no corrosive or powerful solvents are present that may react with seals and containers. Because of the absence of gas formation during operation and of any significant vapor pressure, the battery can be packaged in low-pressure containers such as plastic–metal barrier laminates. Cells can be configured in almost any shape because of the flexibility of the materials involved. Some projected designs are illustrated in Figure 1.3.[2] The actual manufacture of SPE batteries should therefore be easier because of

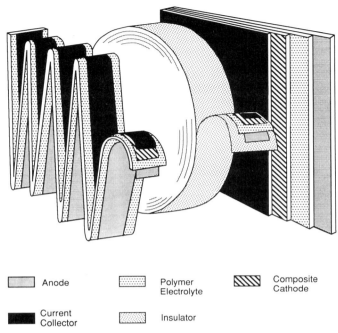

	Anode		Polymer Electrolyte		Composite Cathode
	Current Collector		Insulator		

Figure 1.3. Various configurations for extended-area polymer electrolyte batteries. From Ref. 2.

the elimination of the liquid component; processes may be highly automated, and as new technology is not required—several existing plastic film manufacturing techniques can be used or adapted—the costs of implementing a new approach to storage battery construction are very favorable.

1.3. Performance Characteristics of Laboratory Cells

Passeniemi and Inganäs[3] have developed a model to analyze theoretical performance limits of secondary polymer batteries of the form

$$P_2^- Li^+ \mid PEO\text{–}LiClO_4 \mid P_1^+ ClO_4^- \tag{1.1a}$$

and

$$Li \mid PEO\text{–}LiClO_4 \mid P_1^+ ClO_4^- \tag{1.1b}$$

where P_1 and P_2 are conducting polymers capable of being intercalated or doped with anions and cations, respectively. P_1 is typically polythiophene, polyaniline, or polypyrrole, whereas P_2 is polyacetylene. The model uses the following facts:

1. On charging, salt is consumed and the electrolyte is diluted. On discharging, a reverse process enriches the electrolyte with salt.

2. Polymer electrodes have a maximum intercalation capacity.
3. Electrolytes have an optimum salt concentration.

The cell reactions are

$$Li + n[P_1^+(ClO_4^-)_{1/n}] \underset{charge}{\overset{discharge}{\rightleftarrows}} LiClO_4 + nP_1$$

$$m[P_2^-(Li^+)_{1/m}] + n[P_1^+(ClO_4^-)_{1/n}] \underset{charge}{\overset{discharge}{\rightleftarrows}} LiClO_4 + mP_2 + nP_1$$

$$(1/x - 1/y)LiClO_4 + 1/y[P(EO)_y LiClO_4] \underset{charge}{\overset{discharge}{\rightleftarrows}} 1/x[P(EO)_x LiClO_4$$

The $1/n$ and $1/m$ are the maximum intercalation levels of P_1 and P_2, respectively, and x and y denote the ether oxygen-to-lithium (O:Li) molar ratio at maximum and minimum concentrations (the range of concentrations where the conductivity is maximized). By combining these equations, the necessary material amounts for one Faraday of charge can be derived. Table 1.1 shows the data for polymer battery components necessary to determine theoretical energy densities and weight and volume distributions in polymer batteries that are given in Table 1.2. Volume and weight distributions require alteration if a composite including SPE is used for the electrode. The values in Table 1.2 represent the theoretical maximum values and do not take into consideration the weight and volume of other components of a real battery. It was suggested these values should be divided by a factor of at least 1.5 to get maximum practical values. It was concluded that it should be possible to construct batteries with higher energy density than the Ni–Cd batteries (40 Wh kg^{-1} or 120 Wh dm^{-3}); however, a comparison with the MoliCell, which has a specific energy of 70 to 80 Wh kg^{-1} or 200 to 240 Wh dm^{-3}, implied that a polymer battery could not compete, at least on an energy density basis.

Scrosati[4] has reported calculations of energy density for lithium-based polymer electrolyte batteries utilizing P(EO)$_9$LiCF$_3$SO$_3$ and V$_6$O$_{13}$ cathodes. The calculations were based on practical and optimized thickness and weight estimates of single-cell components and of cell capacity and efficiency for a V$_6$O$_{13}$ positive

Table 1.1. Data for Polymer Battery Components

Component	Material	M (g mol^{-1})	ρ (g cm^{-3})	n	m	x	y
SPE	PEO	44		1.2		8	20
P	PPy	65	1.0	4			
P	PTh	82	1.0	3			
P	PA	13	1.0		5		
MX	LiClO$_4$	107					
M	Li	7	0.5				

Table 1.2. Maximum Theoretical Energy Densities and Weight and Volume Distributions in Polymer Batteries

System	Ah kg^{-1}	Ah dm^{-3}	Cell Voltage on Discharge (V)	Wh kg^{-1}	Wh dm^{-3}
(1a,PPy)	25	29	3.0	78	87
(1a,PTh)	27	30	3.5	93	103
(1b)	24	27	1.5	34	41

	Weight Distribution (%)			Volume Distribution (%)		
	−ve	+ve	Electrolyte	−ve	+ve	Electrolyte
(1a)	1	35	64	2	39	59
(1b)	7	33	60	7	37	57

electrode as given in Table 1.3. The resulting values of energy densities were not particularly high (~80 Wh kg^{-1} for practical, ~200 Wh kg^{-1} for optimized cells operating at 50% efficiency) but this was largely attributed to the high percentage of the cell weight being taken up by nickel current collectors.

Three major groupings have been involved in developing practical large-scale secondary lithium batteries using polymer electrolytes. One based at AERE, Harwell, United Kingdom, has worked on cells with thin-film configurations of various types having a lithium foil anode, an electrolyte such as P(EO)$_9$LiCF$_3$SO$_3$, and a composite cathode of V$_6$O$_{13}$, acetylene black, and polymer electrolyte. Cells were operated in the temperature range 120–140°C. The second group, the Canadian–French ACEP (accumulateurs à électrolyte polymère) Project, has tested cells using TiS$_2$ and V$_6$O$_{13}$ positive electrodes with P(EO)$_8$LiCF$_3$SO$_3$ or P(EO)$_8$LiClO$_4$ electrolyte.[5,6] Current densities of 0.1 to 1.0 mA cm^{-2} have been reported by both groups for PEO–LiCF$_3$SO$_3$ electrolyte cells with energy efficiencies and specific powers of 70 to 80% and 100 to 300 W dm^{-3}, respectively.[6,7] New electrolytes developed by the ACEP group for lower-temperature operation have brought about two generations of batteries. The first development reduced the crystallinity in the polymer by modifying the polymer chain itself, allowing cell operation in the temperature range 50 to 80°C without loss of performance. The second generation involved further chain modification but also the use of a plasticizing lithium salt of the form (CF$_3$SO$_3$)$_2$N$^-$Li$^+$ [1], which has permitted cell operation at ambient temperature. Results from this project have been collated in a review by Gauthier et al.[8]

Table 1.3. Theoretical Energy Densities

Cell Reaction	Wh kg^{-1}	Wh dm^{-3}
Li + TiS$_2$ → LiTiS$_2$	480	1200
8Li + V$_6$O$_{13}$ → Li$_8$V$_6$O$_{13}$	890	2110
Li + V$_2$O$_5$ → LiV$_2$O$_5$	482	1357
3Li + Li$_{1+x}$V$_3$O$_8$ → Li$_4$V$_3$O$_8$	730	2500
Li + MnO$_2$ → LiMnO$_2$	1027	3185
1.45Li + CrO$_{2.86}$ → Li$_{1.45}$CrO$_{2.86}$	1420	2860

Recently, details of a third project, the Mead–H&L (USA and Denmark) Project, have been reported in a series of patents. The electrolyte under investigation is a two-phase interpenetrating network: PEO crosslinked with a polyacrylate as the continuous phase; a dipolar, ion conducting solvent such as propylene carbonate; and an appropriate salt such as $LiCF_3SO_3$.[9,10] The cathode is a V_6O_{13} composite prepared in a particular manner[11,12] (see Section 1.3.3) and a lithium foil anode. The construction of bipolar electrode multicell batteries containing these components has also been described.[13]

1.3.1 Polymer Electrolyte Cells for Operation at $<100°C$

A particular problem in the development of SPE batteries is the poor ambient temperature conductivity of the electrolyte. Many approaches have been reported to overcome this problem (see Chapter 6), but few of these materials have been tested for performance in complete cells. A number of reports have been made on cells incorporating the amorphous polymer poly(methoxyethoxy ethoxyphosphazene) (MEEP). These include $Li_{0.7}WO_2/TiS_2$ cells at $90°C$[14] and Li/TiS_2 cells at various temperatures.[15–18] For the former cells, the open circuit voltage after reaching a steady value during discharge was in the range 2.1 to 2.2 V. Superior lifetimes and electrochemical reversibility were reported when compared with PEO-containing cells. Current densities achieved with $LiWO_2$ as opposed to a Li electrode have been reported to be considerably less.[19] Alamgir et al.[17,18] used a blend of PEO and MEEP and $LiCF_3SO_3$ or $LiClO_4$ as electrolyte to give satisfactory dimensional stability to the system. Significantly improved rate capability relative to PEO-containing cells at $50°C$ was reported. Nazri et al.[15,16] tested cells with MEEP–$LiCF_3SO_3$ and MEEP–$LiClO_4$ electrolyte at ambient temperature. An open circuit voltage of 2.8 to 2.9 V was given for the trifluoromethanesulfonate-based cell. Current densities for cell charging were 30 or 50 μA cm^{-2} with a charge and discharge time of 1 h. One hundred percent current efficiency was reported. Polarization of the electrodes was evident for these cells and occurred during the deintercalation of Li from TiS_2. This polarization, which also limits the Li|PEO–Li salt|TiS_2 cell performance, is due to the fact that the cell potential is largely controlled by the positive composite electrode.[6]

Owens and co-workers[20] have described the fabrication of cells of the form Li|"LiNAGE"|V_6O_{13}, where "LiNAGE" is an unspecified nonaqueous gel electrolyte (polymer, salt, and low-molecular-weight solvent blend) with a room temperature ionic conductivity of 4.5×10^{-4} S cm^{-1}. The open circuit voltage at $22°C$ was 3.6 V, and on constant current discharge at 100 μA, the cell capacity was 0.7 mAh with a cutoff voltage at 1.5 V. Preliminary results indicated good reversibility but interfacial polarization was significant, possibly as a result of blending of $P(EO)_8LiCF_3SO_3$ in the composite cathode. Increased electrochemical performance in terms of cell capacity, coulombic efficiency, cycling energy efficiency, and specific energy on cycling was reported and is under further investigation.

Munshi and Owens[21] investigated the performance of cells of the form Li|$P(EO)_8LiCF_3SO_3$|V_6O_{13} and Li|$P(EO)_8LiCF_3SO_3PC$|V_6O_{13}, where PC is propylene carbonate, in the temperature range 25 to $100°C$. Cells without PC demonstrated an open circuit voltage of >3.5 V at room temperature which reduced to 2.9 to 3.1 V at 100°C. PC cells fabricated for room temperature operation had open circuit voltages of 3.12 to 3.27 V. With 10 vol% PC, an initial capacity of 64% was obtained which declined on cycling: after 20 cycles, the cell capacity reached below 20% of the theoretical level. This decline was even more pronounced at 40 vol% PC. Optimum results were obtained for 20 vol% plasticizer. The initial capacity reached ~86% and the capacity decrease was less rapid, with approximately 26 cycles obtained before significant decline was noted. Room-temperature discharge rates of up to 100 μA cm^{-2} were reported with good cycling and recharge-ability.

New polymer–salt complexes developed for the ACEP project have been used in ambient temperature test cells.[6,8] Typical performances with primary high-voltage MnO$_2$ electrode and the reversible low-voltage MoO$_2$ electrode both show utiliza-tions in excess of 80% at rates up to C/37 (i.e., discharged in 37 h). In addition, nonoptimized MoO$_2$ cells have shown >20 cycles at deep discharge (>75%) at C/36 and 400 cycles at rates between C/50 and C/10 at 27°C. Good electrochemical properties were maintained at C/50, even after 300 cycles at C/10. Other factors relating to this polymer are given in following sections.

1.3.2. Batteries Utilizing Polymer Electrodes

With the continuing advances in semiconducting polymers such as polyacetylene, polypyrrole, and polythiophene, which can be *p*- or *n*-doped electrochemically, there is much interest in developing these polymers as electrode materials.[22–26] A few reports on all-polymer cells have been made. Chiang[27] described a cell based on P(EO)$_5$NaI and polyacetylene,

$$(CH)_x \mid P(EO)_5NaI \mid (CH)_x$$

which, when charged by the passage of *xy* Faradays, could be written as

$$(CHNa_y)_x \mid P(EO)_5NaI \mid (CHI_y)_x$$

Such cells could be cycled for about 20 times at low rates. Nagatomo et al. studied a similar cell using poly(vinylidene fluoride)–LiClO$_4$, plasticized by PC as the elec-trolyte.[28,29] An experimental energy density of 47 Wh kg^{-1} was obtained. No change in coulombic and power efficiencies of the battery was observed even after 300 charge–discharge cycles, although (CH)$_x$ doping was reported to be low.

Scrosati[30] has reported some results on the cell Li|PEO–LiCF$_3$SO$_3$|polypyrrole. The open circuit voltage at 100°C is 3.15 V. The operating capabilities of the battery show good cyclability. Arbizzani and Mastragostino[31] have studied the cyclability features at 70°C of the cell Li|P(EO)$_{20}$LiClO$_4$|pBT, where pBT is polybithiophene. An initial activation phase and subsequent decrease in capacity with the cycle number are evident for this cell and for a similar one using a liquid electrolyte. Cycles 30 to 300 gave efficiency values of 98 to 99%, whereas for the first few cycles this was low. Self-discharge rates were found to be high at 70°C but not appreciable at room temperature. The origin of the high-temperature discharge was unknown but did not involve electrode or electrolyte degradation.

1.3.3. Rate and Cycle Capability

A phenomenon that is often found in the performance of intercalation cathode-based rechargeable batteries is that of capacity decline with cycling. Figure 1.4 shows typical first discharge curves for small V_6O_{13}-based cells containing PEO–$LiCF_3SO_3$ electrolyte at different current densities. The well-defined structure of the curve, associated with ordering effects during the insertion process, is indicative of rapid kinetics. On recharge, there is a general loss of structure in the cell voltage curve, accompanied by a fall in capacity over the initial discharge. The magnitude depends on the current density as well as cathode composition and the value of the upper voltage limit. Typically, for a charge current of 0.15 mA cm^{-2} to a limit of 3 V, the loss is approximately 10 to 20%. Figure 1.5 shows that high utilizations can be maintained at higher rates of discharge. In many cases, however, the rate of decline is low, especially after the first few cycles. The reason for this is not fully understood although it appears to be associated primarily with the composite cathode. Factors such as morphological changes, kinetic limitations, and irreversible insertion of lithium into the crystal structure may all contribute. Similar effects were noted by Bonino et al.[32] for cells using $Li_{1+x}V_3O_8$ intercalation electrode. These authors stressed that as this particular cathodic material can accept up to at least 3 Li^+ mol^{-1} without undergoing structural alterations (when it did, reversibility was lost), the capacity decline was therefore more likely to result from macroscopic factors such as deformation and contact losses rather than irreversible structural changes in the intercalation host.

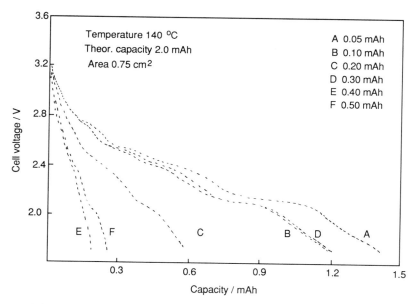

Figure 1.4. Initial discharge characteristics of Li–V_6O_{13} cells incorporating a PEO–$LiCF_3SO_3$ electrolyte at different current densities. From Ref. 7.

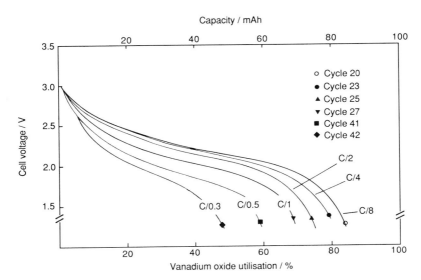

Figure 1.5. Rate performance with cycling for Li–V_6O_{13} cells incorporating a PEO–LiCF$_3$SO$_3$ electrolyte. From Ref. 41. This paper was originally presented at the Fall 1988 Meeting of The Electrochemical Society, held in Chicago, Illinois.

The performance of cells in terms of current densities and initial capacities varies depending on the fabrication of the cathode. Different methods give different levels of inhomogeneities and this has been discussed by Hooper and North.[7] Performance characteristics of polymer electrolyte cells have been reported to be dramatically increased by the use of a novel V_6O_{13} cathode construction.[11,12] This involves formation of a layer (20–25 μm) of polymer electrolyte and activated carbon encapsulating vanadium oxide spheres by forming an emulsion with a suitable organic solvent and depositing it as a thin film on the current collector. This improves the surface area contact and it is claimed that this preparation method leads to increased cell lifetime and performance. A cell using sodium thiochromite composite cathode, Li|P(EO)$_8$LiClO$_4$|Na$_{0.1}$Cr$_{0.8}$V$_{0.2}$S$_2$, has been studied by Geronov et al.[33] A specific energy of 100 Wh kg^{-1} was calculated for the cell. At 140°C, between 50 and 80% utilization with respect to the theoretical capacity was achieved over a wide current density range. At 115°C this dropped to only 30% at 0.27 mA cm^{-2}. This is only half the value reported by Gauthier et al.[6] for TiS$_2$; however, values at 140°C are better than reported by Hooper and North for V_6O_{13}.[7] Utilization of the active cathode material was also reported to be superior over the first 40 cycles when compared with V_6O_{13}.

The behavior of first- and second-generation ACEP cells as a function of discharge rate is given in Figure 1.6[34] for TiS$_2$ and MoO$_2$. The drastic decrease in utilization with only small current density increase is expected for transport-controlled limited electrolytes and is typical of first-generation power curves. Greatly improved performance is found for second-generation cells. Also illustrated is the effect of increased capacity for TiS$_2$.

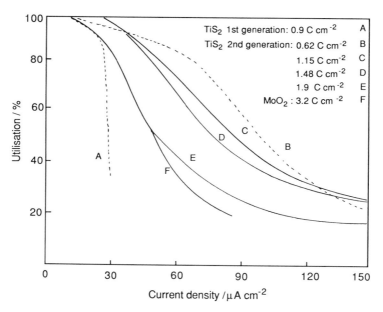

Figure 1.6. Improved cell performance as a function of current density and capacity per unit surface. From Ref. 34. This paper was originally presented at the Fall 1987 Meeting of The Electrochemical Society, held in Honolulu, Hawaii.

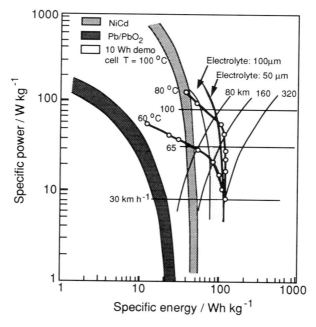

Figure 1.7. Ragone plot for small cells at 60 to 80°C for a 10-Wh demonstration cell at 100°C with electrolyte thicknesses of 50 and 100 μm. The km-km h^{-1} grid as defined by Ragone[40] is also represented. From Ref. 8.

Cell cycling performances indicate that SPE lithium batteries can be fully discharged for over 500 cycles at high to moderate temperatures.[8,35] With second-generation ACEP cells having improved ambient temperature operation, deep discharge cycling in excess of 700 cycles has been reported with TiS_2 cathodes.[34] As no observable memory effects have ever been reported for SPE cells, it may be assumed that much higher cycling values would be found for shallow discharges.

The SPE concept meets most of the requirements for the optimization of lithium cyclability, that is, low surface charge density, use of low charging currents, stable and well-defined elastomeric interfaces, and good electrolyte–lithium stability. Lithium or lithium alloys are known to form passivating films at the electrode–electrolyte interface. Such films are discussed more fully in Chapter 10. These films may in fact contribute to electrodeposited lithium "recontacting," and the idea has been used by some authors to explain the high efficiency of lithium in the presence of solid electrolyte films.[36,37] At any rate, the film does not in general hinder the ionic and electronic exchanges between lithium and polymer electrolyte.

Expressing lithium rechargeability in terms of a figure of merit (FOM), where FOM = number of cycles × average capacity stripped/total capacity available, the values for liquid organic electrolyte systems rarely exceed 30 to 50. The beneficial effects of low surface capacities, effectively dendrite-free lithium electrodeposition, lack of co-insertion and penetration of solvent into the electrode materials, and polymer flexibility and adhesiveness add up to give FOM values that can be in excess of 120 for these cells.[38] Values of approximately 50 have been achieved for bipolar configurations.

1.3.4. Dendrite Formation

Polymer electrolyte technology is a surface technology requiring low (<4 mAh cm^{-2}) surface capacities. Consequently, the current densities are always low and, as a result, favor electrochemical redeposition of lithium without dendrites. The occurrence of lithium dendrites in SPE cells has been reported in early work on batteries.[6] Better control of technical parameters has now practically eliminated this reason for cell failure. These dendrites tend to appear toward the end of a charge and have been interpreted as an electronic leakage current in competition with the ionic current.[8] Armand[39] reported that other phenomena may well contribute to the inhibition of dendrite growth. These include the mechanical properties of the electrolyte, which may help to push back the dendrites toward the interface, and low current densities, which allow surface reequilibration from self-diffusion of the metal and the chemical inertness of the polymer itself.

1.3.5. Power–Energy Relationship

The Ragone plot is a useful representation for comparing the key power and energy characteristics of different battery systems.[40] In Figures 1.7 to 1.9 the behavior of two types of polymer electrolyte battery, one with a configuration optimized for electric traction applications and the other corresponding to small consumer-type applications. Figure 1.7 represents results for PEO–LiClO$_4$ cells at 95 to 100°C together with the second-generation ACEP polymer electrolyte at various temperatures. Typical calculated specific power and specific energy requirements for steady driving are also shown. The results are impressive although an objective of $\sim$150 Wh kg^{-1} set by the ACEP Project groups requires further refinements to be reached. Figures 1.8 and 1.9 show the volumetric energy and power capacities. Lines of constant discharge times are also included. Effects of electrode surface loading, operating temperature, and SPE formulation are illustrated. With the power characteristics significantly upgraded at room temperature by the use of improved electrolyte, it is expected that further optimization will enhance the volumetric energy content and reduce operational temperatures further, as shown in Figure 1.9.

1.3.6. Self-Discharge

Self-discharge in secondary batteries is usually a result of decomposition caused by water or electrode material solubility or reaction with the electrolyte. In theory, in

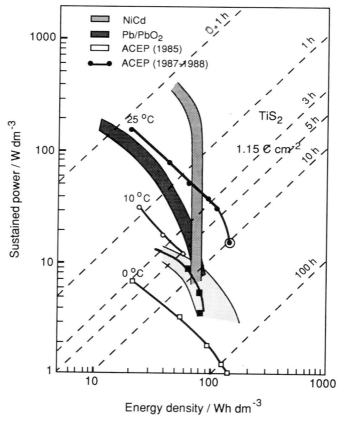

Figure 1.8. Ragone plot for a second-generation SPE TiS$_2$ cell (1.15 C cm^{-2}) at 25, 10, and 0°C. A comparison is made with a first-generation electrolyte at 25°C (ACEP 1985). Lines of constant discharge times are also shown. From Ref. 8.

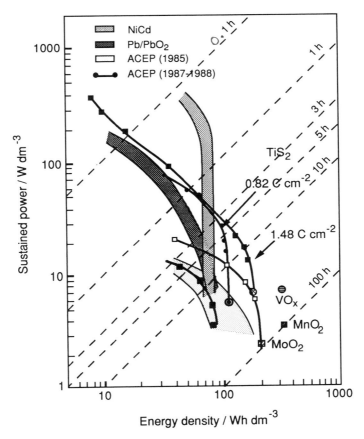

Figure 1.9. Ragone plot for second-generation SPE cells showing the effect of electrode surface loading for TiS_2 (0.82 and 1.48 C cm^{-2}) and MoO_2 (3.2 C cm^{-2}). Partial results on VO_x and MnO_2 illustrate higher volumetric-energy contents. From Ref. 8.

polymer electrolyte batteries no secondary reactions are possible that would lead to self-discharge. For cells with insertion cathodes, the open circuit voltage can provide a good indication of self-discharge. For lithium SPEs the following self-discharge figures have been measured[8,34,41]:

<0.01% per annum for TiS_2 and MnO_2 at 25°C

~0.15, 3, and 8% for MnO_2, VOx, and TiS_2 per annum at 60°C

<4.0% per annum at 95°C

<20% per annum at 106°C

Above 60°C, some chemical process may contribute to the self-discharge rate. The preceding values are much lower than equivalent values for nickel–cadmium and lead–acid systems and are favorable for long life expectancies during cycling and for a good shelf life. At the highest temperatures, the loss is equivalent to

Lead-acid battery

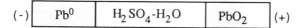

$(-)$ | Pb^0 | $H_2SO_4\text{-}H_2O$ | PbO_2 | $(+)$

Electrolyte participation in electrode reactions.

Overall reaction: $Pb + PbO_2 + 2H_2SO_4 \rightleftharpoons 2PbSO_4 + 2H_2O$

Energy density (Wh kg^{-1}): 250 (theoretical), ~25-35 (practical, ~1/8).

Secondary reaction: $H_2O \rightarrow H_2 + \frac{1}{2}O_2$

%Efficiency (Ah and Wh): 85 and 70-75%.

Cycling characteristic (full discharges): short cycle life (300 cycles).

Operating temperature range: -20 to 40°C (-40 °C for SLI use).

Nickel-cadmium battery

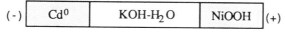

$(-)$ | Cd^0 | $KOH\text{-}H_2O$ | $NiOOH$ | $(+)$

Water participation in electrode reactions.

Overall reaction: $Cd + 2NiOOH + 2H_2O \rightleftharpoons 2Ni(OH_2) + Cd(OH_2)$

Energy density (Wh kg^{-1}): 245 (theoretical), 35 (practical, ~1/7).

Secondary reaction: $H_2O \rightarrow$ ~~~~~~

% Efficiency (Ah and Wh): 7~~~~

Cycling characteristic: good ~~~~~~les), memory effect.

Operating temperature range~~~~

Typo ?!!

SPE battery

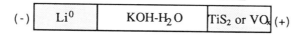

$(-)$ | Li^0 | $KOH\text{-}H_2O$ | TiS_2 or VO_x | $(+)$

SPE is simply a Li$^+$ ion carrier that can be made ultra-thin.

Overall reaction: $TiS_2 + x\,Li^0 \rightleftharpoons Li_xTiS_2$

Energy density (Wh kg^{-1}): 480 (theoretical), ~120 (practical, ~1/4).

No secondary reaction.

%Efficiency (Ah and Wh): ~100 and 85-95%.

Cycling characteristic: full discharge capability (600 cycles).

Operating temperature range: -10 to 130°C.

Figure 1.10. Main differences between SPE lithium batteries and existing aqueous systems. From Ref. 8.

1 mV day^{-1} or a shelf life of 4 to 5 years. Storage at room temperature suggests a shelf life of 20 to 25 years.

1.3.7. Overcharge Behavior

Corrosion and cell failure resulting from the overcharging of V_6O_{13}-based cells for long periods (>30 min) have been reported by Owens and co-workers.[42] This was due to degradation of both the electrolyte and the cathode at approximately 5 V. Lesser voltages can, however, be withstood for considerable times.

To summarize, Figure 1.10 shows the main characteristics and differences between three battery technologies: lead–acid, nickel–cadmium, and SPE.

1.4. Scaled Up Batteries

Several important issues need to be addressed before batteries can be manufactured. These include cost, fabrication techniques, long-term electrochemical behavior, and alternative cell assemblies; cells can be stacked in series or parallel to increase the voltage or current, respectively. One approach suggested by the Harwell group is to use bipolar modules,[43] which would reduce the weight and cost during scaleup. Such a system is represented in Figure 1.11. Although unipolar electrode design is a simpler approach, a number of disadvantages exist. One is the long current path

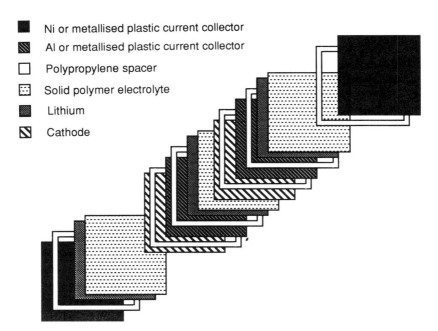

Figure 1.11. Schematic diagram of a prismatic bipolar (2 units) battery. From Ref. 44.

along the thin current collector of one electrode perpendicular to the direction of electrolytic current, through the intercell connector, and then along the thin current collector of the opposite electrode of the next cell. The resulting ohmic loss not only has an adverse effect on limiting the power density and reducing the voltage efficiency, but results in the generation of excess heat. In addition, this design may lead to nonuniform current distribution and, on charge and discharge, may result in poor utilization of the active materi₂l and affect cycle life. The use of bipolar electrodes minimizes the length of the current path between cells connected in series; that is, electrons travel from the electrode material of one polarity in one cell compartment to the electrode material of opposite polarity in the adjacent cell. This design eliminates the need for an additional intercell connection with the result that ohmic loss is reduced considerably and is independent of the cell area or configuration. Munshi and Owens[44] have assessed the theoretical energy density performance limits for packaged rechargeable lithium polymer electrolyte batteries, based on $P(EO)_8$–Li salt (conductivity equal to 10^{-4} S cm^{-1}) and a composite V_6O_{13} (75.4 wt % V_6O_{13}, 5.0 wt % carbon, and 19.6 wt % electrolyte) cathode of both unipolar and bipolar electrode configuration. For a unit cell, an optimum specific energy of 240 Wh kg^{-1} is calculated for cells with nickel current collectors, a 100-μm-thick cathode, and 100-cm^2 cell areas, whereas for metallized plastic current collectors, 460 Wh kg^{-1} is calculated for a 100-μm-thick cathode with the same cell area. Further small increases in the energy densities are obtained with the bipolar configuration. An optimum specific energy of about 450 Wh kg^{-1} is obtained for cells using an aluminum current collector with 20 bipolar units, 400-cm^2 cell area, and 100-μm-thick cathode. By the use of metallized plastic current collectors, only a small increase in the specific energy to about 470 Wh kg^{-1} for cells with 5 bipolar units, 100-μm-thick cathode, and 400-cm^2 cell area is found. An optimum cell area of between 100 and 400 cm^2 is found for all cases. The results yield an upper limit for performance capability that is in excellent agreement with work carried out by Hooper and co-workers.[41,43]

Although a Ragone plot is a useful tool to describe a battery's performance and behavior, there is, however, a correlation between the cell performance and loading capacity of the cathode where the full cell reaction takes place. Cells with higher capacities (thicker cathodes) will yield higher energies at low power with the added benefit of a relative reduction in weight in the active components. On the other hand, cells with low capacities (thinner cathodes) will yield higher powers and hence higher currents with correspondingly lower energies. Hence the configuration of a particular battery system depends on the type of application. Munshi and Owens further described the theoretical aspects of SPEs by assessing the specific energy and power of unipolar systems with two types of current collector and different loadings of the positive electrode.[45] Figure 1.12 shows Ragone plots for different battery systems, comparing calculated values for 20-μm-thick electrolytes, 100-μm-thick cathode cells using nickel current collectors and cell areas ranging between 25 and 400 cm^2 with practical values for nickel–cadmium, lead–acid, and polymer electrolyte 10-Wh cells having an electrolyte thickness of 50 μm and operating at 100°C.[8] A specific power higher than that obtained for the practical

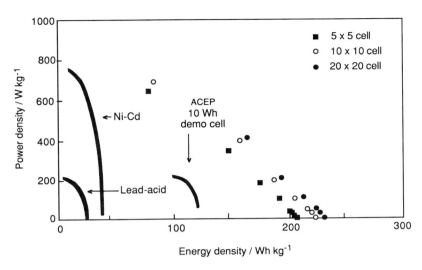

Figure 1.12. Ragone plots for different battery systems. From Ref. 45.

demonstration cell is found; however, calculated power values include only ohmic losses and thus are higher. Despite this, the calculations indicate the operating range of the system. Use of bipolar configurations should further increase the specific energy and power. Further assessments by these authors have involved modeling of SPE batteries of various configurations capable of delivering high specific pulse powers.[46]

The performance of a scaled up battery must be similar to that of small-scale laboratory prototypes in all respects, from electrochemical performance to fabrication technique. For example, it is important that individual cells stacked in various arrays are well balanced in capacity to avoid overcharging or overdischarging. This is a critical point with polymer electrolyte-based batteries as there are no side reactions to provide a cell leveling mechanism in this system. A 10-Wh high-temperature demonstration cell based on commercial PEO and a 1-Wh low-temperature cell incorporating modified PEO have been built by the ACEP group.[1] Surfaces are approximately 1000 times larger than those of the usual test cells. No unfavorable scale effects were observed in the performances, including cycle tests. Although these results are encouraging it is still insufficient for assessing the influence of cumulative volume variations or the thermal management of larger-scale devices.

A basic disadvantage of a polymer electrolyte battery is that the weight and volume of essential but nonreactive components significantly reduce the achievable energy and power densities.[41,44] These may be minimized by choice of appropriate materials and fabrication. Several materials can be used to increase the energy and power density of this system. They include low-density metals (e.g., aluminum), metallized plastic current collectors, and electronically conductive polymers.

Thermal management is necessary for all electric vehicle battery systems, whatever the optimum operating temperature, as the temperature ranges are affected by internal heat generation and external cooling air. For SPE batteries, the main factors controlling thermal management combine to give very favorable conditions.[8] For example, high-energy efficiency results in little waste heat and the substantial heat capacity aids temperature stability. A wide ($\sim$50°C) optimum operating temperature bandwidth has already been established for amorphous polymer electrolytes and is likely to be improved further in the future. In addition, the cell operating temperature (60–80°C) is sufficiently higher than the maximum expected ambient temperature ($\sim$40°C) to allow rapid heat removal, when required, but not so high as to necessitate the use of high-performance thermal insulation or major external heating for low-temperature (-20°C) operation.

Thermal modeling studies have been carried out at Harwell to obtain information about the potential thermal behavior of practical cells and batteries as a function of size, geometry, internal design features, insulation, and duty cycle.[41] Small internally generated temperature rises are predicted for batteries operating at rates consistent with vehicle traction applications. Thermal management problems are expected for the elevated temperature system when it is operating at low or zero discharge rates. An example has been given of a 21.2-kWh (424-Ah, 50-V) battery that was assumed to consist of 40 cells of 15 × 15 cm, each with a capacity of 212 Ah and a nominal average operating voltage of 2.5 V. When discharged at a mean rate of C/10, the temperature could be maintained at 130°C with 66-mm thickness of insulation, whereas at a rate of C/2.5, the same temperature could be maintained without any insulation. The battery could run for 1 h at a discharge rate of C/1.85 before the mean temperature rose by more than 35°C. An operating temperature range of 100 to 140°C was concluded to be quite acceptable.

Gauthier et al.[8] have described an evaluation of an SPE power source carried out by the ACEP group for electric propulsion, using the urban van as an illustration. The experimental results from studies on small-scale SPE lithium batteries suggest that, in principle, application of such power sources to this usage is indeed feasible. The choice of cell for the design concept was a large parallel-wound 280-Wh single cell module rather than the bipolar modules proposed by the Harwell group.[7] The guidelines for the design were

weight $\leq$ 700 kg

volume $\leq$ 600 dm^3

power availability at all times $\geq$ kW

energy capacity $\geq$ 21 kWh

system voltage = 120–240 V

battery life $\geq$ 500 cycles or 2 days

The proposed power source comprised 144 cell modules, that is, a 40-kWh nominal capacity and a voltage range of 120 to 240 V. The overall power source volume and weight were given as 505 dm^3 and 407 kg.

Despite progress in terms of the number of cycles, operating temperature, and energy density, which make SPE batteries closer to fulfilling all requirements for viable application, further work on cell aging behavior is required; the technology has yet to be demonstrated at full scale and with this, the cost optimization must be favorable toward the new processing technology.

1.5. Electrochromic Devices

1.5.1. Electrochromic Materials

Electrochromism may be broadly defined as the phenomenon that gives rise to a reversible color change, brought about either by the electrochemical insertion and extraction of electrons and ions into inorganic materials, for example, tungsten oxide, WO_3, or organic materials such as diphthalocyanines[47] and tetrathiafulvalenes,[48] or by an electroredox reaction in an organic material. Viologen[49] is an example of this latter class. An electrochromic device is therefore charge controlled, unlike liquid crystal displays (LCDs), for example, which are electric field dependent. During the last two decades, there has been a growing interest in electrochromic (EC) materials for use in practical electrooptical devices: EC displays, optical modulators, adjustable reflectance mirrors for vehicles, sunglasses, and smart window glass for control of energy transfer within buildings and in vehicles.

The first reports of electrochromism, in thin-film tungsten oxide and molybdenum oxide, were made by Deb[50] in 1969. Practical devices, which gave color switching times of less than 1 s, were fabricated in the early 1970s[51] but device failure on repeated cycling, caused by side reactions between the aqueous electrolyte and the inorganic electrochromic material, limited their operating life. The use of nonaqueous electrolytes, in particular lithium perchlorate–propylene carbonate with WO_3, overcame this particular problem[52] and has been the most commercially suitable arrangement and commanded the most extensive studies.

Electrochromic materials can be divided into two classes. The first group comprises those that color cathodically by a reduction process, such as the oxides of W, Mo, V, Nb, and Ti,[52] viologen,[49] and a number of conducting polymers[53,54] and gels.[55] The second group, which colors by an oxidative process, includes oxides of Ni, Rh, Co, and Ir[56–58] and Prussian blue.[59] The properties of various electrochromic materials are well documented[52,56,57] and have been investigated for their suitability in electrochromic devices. Although many, including mixed oxide systems,[58] show specific characteristics that are superior to those of WO_3, it is presently the most suitable device material because of its reliability over extensive cycling and relative ease of thin-film deposition.

Coloration in tungsten oxide is induced by the electrochemical reaction

$$xM^+ + WO_3 \rightleftharpoons M_xWO_3 \qquad \text{where, typically, } M^+ = Li^+, Na^+, H^+$$

and by altering the applied voltage, the transmittance of the system can be varied. The coloration is considered to originate from the resonance transfer of trapped electrons between equivalent tungsten ions. It is important to note that, unlike LCDs, EC reactions exhibit open circuit memory; that is, a reverse polarization is required to restore the colorless state.

1.5.2. Electrochromic Displays

An electrochromic display is a multilayer thin-film electrochemical cell. By applying a voltage between the electrochromic layer and the counter electrode, which are separated by a transparent ionic conductor, the electrochromic reaction can be driven. Displays operate in a diffuse reflectance mode, and the requirement of the counterelectrode is simply to provide the electrochemical balance. Lithium or sodium metal would be appropriate counterelectrodes for this purpose. Figure 1.13 shows a schematic structure of a solid-state electrochromic display using a polymer electrolyte.

1.5.3. Electrochromic Windows

A typical multilayer arrangement for smart window applications differs from the previous display configuration in that, unlike electrochromic displays that operate in a reflectance mode, the entire window is in the optical path and therefore the counterelectrode in this instance must either be (1) optically passive, that is, colorless whether in the oxidized or reduced state; or (2) electrochromic also but in a complementary sense. An example of the latter option is shown in Figure 1.14. The former mode has been the more difficult to achieve but the advantages of the latter, for example, potentially lower applied voltage and power consumption requirements to bring about a specific optical density change, make it particularly appealing.

1.5.4. All-Solid-State Devices

The problems associated with liquid electrolytes in devices center around the fabrication of large-surface-area displays. For example, a liquid-filled gap of the order of

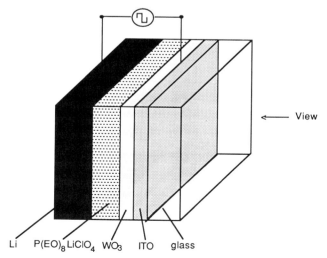

Figure 1.13. Schematic diagram of a solid-state electrochromic display using a polymer electrolyte.

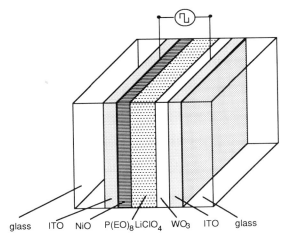

Figure 1.14. Schematic diagram of an electrochromic window where the two electrodes are electrochromic but in a complementary sense.

glass   ITO   NiO $P(EO)_8LiClO_4$ WO_3 ITO   glass

100 μm between glass sheets of a few millimeters in thickness and tens of square meters in area will readily deform. A solid electrolyte offers a promising alternative technology. Several all-solid-state devices have been reported where thin films of materials such as $LiAlF_4$,[60] $LiNbO_3$,[61–63] $LiTaO_3$,[61–63] Li_3N,[64] $Na^+-\beta-Al_2O_3$,[65] and NASICON[66] have been incorporated as the ionically conducting layer. Although prototype smart windows have been successfully fabricated,[63] important problems relating in particular to deposition of layers, susceptibility to mechanical stress, and device failure still remain to be resolved.

1.5.5. Polymer Electrolytes in Electrochromic Devices

As in rechargeable battery applications, deposition of a polymer electrolyte thin film that is both solid and flexible can eliminate many of the limiting factors associated with either liquid or rigid solid ionic conductors. The advantages of a polymer-based electrochromic device are many: film deposition of large surface areas is straightforward, with the necessity to sputter-deposit glassy films[67] eliminated along with its inherent problems. In addition, the flexibility of the material makes it mechanically robust and, overall, well suited to large-area-device fabrication.

1.5.5.1. Response Time

For all forms of electrochromic devices, the response time to switch the optical state must be suited to the device's requirements. Goldner[68] suggests appropriate switching times of less than 100 s for building windows, less than 1 s for vehicle mirrors and windows, and 1 s to less than 1 ms, depending on the particular application, for displays.

The temperature dependence of the response time of a display incorporating

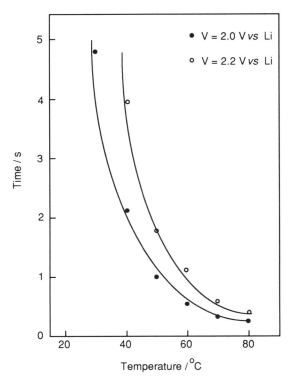

Figure 1.15. Effect of temperature on response time for coloration to obtain an optical density variation of 0.3. From Ref. 69.

$P(EO)_8LiClO_4$ electrolyte[69] is shown in Figure 1.15. This is for an optical density variation of 0.3 and for applied potentials of 2.0 and 2.2 V. For a potential of 2.0 V, coloration times of 450 and 250 ms were obtained at 60 and 80°C, respectively. The bleaching response time was 600 ms at 60°C and 170 ms at 80°C for a 3.6-V applied potential. Although at 25°C, the kinetics are very slow, of the order of tens of seconds, it was demonstrated not to be necessary to have a completely amorphous system to obtain reasonable rates of response.

1.5.5.2. Open Circuit Memory

A persistence of open circuit memory is an essential requisite in EC displays. Although liquid-based cells usually exhibit a faster response time, solid cells have been found to have superior open circuit memory.[57] In all EC displays that use a polymer electrolyte, this parameter has been reported to remain constant over long periods. Kobayashi[70] observed some 10% loss in optical density over the first 3 h at open circuit, which was attributed to a slow diffusion of Li^+ ions at the WO_3- polymer electrolyte boundary to compensate the local charge resulting from the perchlorate counterions.

Scrosati and co-workers have reported preliminary data on an electrochromic window.

$$Li_xWO_3 \mid P(EO)_8LiClO_4 \mid NiO$$

that uses complementary coloring and bleaching.[71] WO_3, which can be prelithiated, is colored in this form whereas NiO, which also undergoes reversible lithium intercalation, bleaches with ion insertion. Figure 1.16 gives typical voltammetric curves for this cell at 40, 60, and 80°C. In the cathodic cycle the cell is transparent, and in the anodic cycle the cell is dark, with absorption caused by both NiO and WO_3. The temperature is critical for this cell as ionic conductivity of PEO systems is too low for significant electrochromism below 60°C. The time response of the current and cycling durability are shown in Figure 1.17. Both coloration and bleaching times are approximately 10 s. Electrochromism diminished after about 1000 cycles because of irreversible reactions at the electrochromic/electrolyte interfaces.

1.5.5.3. Thermoelectrochromic Displays

High-molecular-weight PEO-based polymer electrolytes show low room temperature conductivity (10^{-8} S cm^{-1}) because of the crystalline component of the

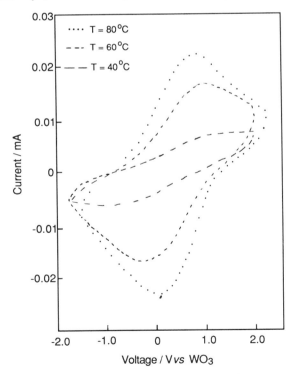

Figure 1.16. Cyclic voltammogram for a $NiO|P(EO)_8LiClO_4|Li_xWO_3$ cell at three temperatures. The voltage scan was 10 mV s^{-1}. From Ref. 71. Reprinted by permission of the publisher, The Electrochemical Society, Inc.

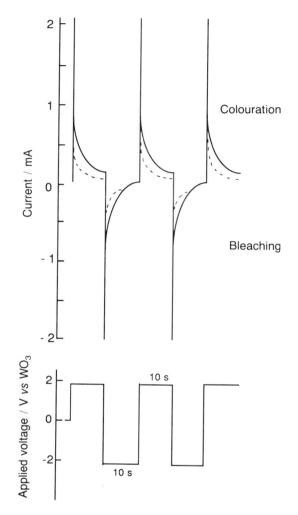

Figure 1.17. Current versus time for a NiO|P(EO)$_8$LiClO$_4$|Li$_x$WO$_3$ cell at 80°C subjected to voltage pulses. Results are for initial cycle (—) and after approximately 1000 cycles (---). From Ref. 71. Reprinted by permission of the publisher, The Electrochemical Society, Inc.

material. At first this may appear to limit its usefulness in electrochromic devices but in fact, it may be exploited in display areas that could not be readily achieved using a liquid electrolyte. The cyclic voltammograms of an electrochromic display using WO$_3$ and PEO–LiClO$_4$ as electrolyte show that, as the temperature is increased to above the PEO melting point at 65°C, one moves from a state of virtually no electrochromic effect to one of sharp contrast. The thermal dependence of the response time shows analogous behavior: between 50 and 70°C, the chromic response increases tenfold, from 20 to 2 s. There are a number of ways of exploiting such a critical thermal response, for example, as a thermoelectrochromic display or, alternatively, as an optoelectrochromic display where illumination induces the required heating effect. Figure 1.18 shows a display of this sort fabricated by Scrosati and co-workers.[72] A laser beam induces an electrochromic response in a confined area of the display. Because of the open circuit memory effect display writing is possible, which can be wiped by altering the temperature of the specific spots. The temperature at which electrochromic response occurs and the range of temperatures over which response does or does not occur may be varied by changing the composition of the polymer electrolyte.[73]

Armand et al.[74-76] have proposed a similar type of display that differs in that it relies on a transition through the glass transition temperature rather than a crystalline-to-amorphous transition. This eliminates problems in the clearing process that may arise through slow recrystallization. In this instance, a protonic conductor, polyvinyl pyrrolidone (PVP):H_3PO_4, was used. For an O:H concentration of 0.2, the conductivity differential on increasing temperature is greatest and the response time for this polymer electrolyte EC display shows near-vertical response at high temperature. Concentration cell effects and lateral diffusion of ions were found to be very slow in a test 1-mm colored pinpoint within a display. No noticeable color diffusion was detectable over 3 months, and applications requiring long-term memory effects and stability of areas within a display are thus feasible.

1.5.5.4. Polymer Electrolytes Other Than PEO

For normal display purposes or in smart windows, high-molecular-weight PEO has limited application. Kobayashi[70] has carried out extensive studies of electrochromic devices using an all-amorphous polymer electrolyte, an oligo(oxyethylene) methacrylate $CH_2{=}C(CH_3)COO(CH_2CH_2O)_7{-}CH_3$, ($MEO_7$), with $LiClO_4$ as the dis-

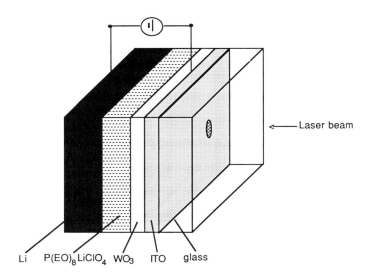

Li $P(EO)_8 LiClO_4$ WO_3 ITO glass

Figure 1.18. Schematic diagram of a thermoelectrochromic display.

solved salt. An EO:Li ratio of 16:1 (6 mol%) gave a room temperature conductivity of 10^{-6} S cm^{-1}. A response time of the order of 2.5 s at room temperature to reach a saturation optical density at applied potentials of approximately 1.2 V compares well with the high-temperature response of a PEO-based electrolyte. Wixwat et al.[77] have studied optical properties of electrochromic smart window-type displays using the polymer blend PMMA–PPG(4000)–LiClO$_4$ as electrolyte. This has a room temperature conductivity of 2×10^{-6} S cm^{-1}. Light transmittance is approximately 89% for a 25-μm-thick electrolyte layer and shows low optical scattering.

1.6. Photoelectrochemical Cells

Photoelectrochemical cells are photovoltaic devices based on the junction of a semiconductor photoelectrode and an electrolyte containing a redox couple, often of the type $I^-/I_3^-/I_2$ or $S^{2-}/S_x^{2-}/S_{x+1}^{2-}$.[78] These systems generally employ n-type semiconductors and transparent electrodes. Illumination produces photogenerated holes that oxidize the complex anions in solution, while electrons travel from the semiconductor through an external load to reduce the species at the transparent illuminated surface. Rediffusion in the electrolyte of the species completes the electrochemical cycle.

Photocorrosion and side reactions at the semiconductor interface involving the electrolytic solvent limit the working life of conventional solution photoelectrochemical cells. Polymer electrolytes offer a means of eliminating this problem. They also allow the electrolyte thickness to be reduced, improving transparency. Such a cell is represented in Figure 1.19. Skotheim and co-workers[79–82] have studied cells of the form

n-Si | PEO–KI–I$_2$(8:1:0.1) | polypyrrole | Pt
n-Si | PEO–KI–I$_2$(8:1:0.1) | ITO

These photoelectrochemical cells gave photovoltages greater than 0.4 V. The photocurrents obtained with optimized electrocatalytic silicon surfaces were limited by the series resistance of the cell. This should be significantly reduced if polymer electrolytes of higher conductivity are used. A second type of cell was proposed by Sammels and Ang,[83] who formed a cell using a single crystal p-indium phosphide photocathode. The electrolyte/redox system consisted of PEO–NaSCN–NaS$_2$/S (4.5:1:0.1). Once again, electrolyte resistance limited the current. Marsan et al.[84,85] have studied all-solid-state polymer electrolyte photoelectrochemical cells

n-CdSe | modified PEO–M$_2$S | xS | ITO

where $1 \leq x \leq 7$ and the modified PEO was of a reduced crystalline form and therefore had improved ionic conductivity. It was also suggested that as the characteristics of cells using low- and high-molecular-weight PEO are very similar,[86] a liquid system, which has a much superior conductivity, may be used to study the behavior of polymer-based photoelectrochemical cells. It has been shown[84] that the conversion efficiency of the cell does not depend on the S:Na$_2$S ratio (x value) but is a function of the nature of the alkali metal cation M^+, increasing from lithium to

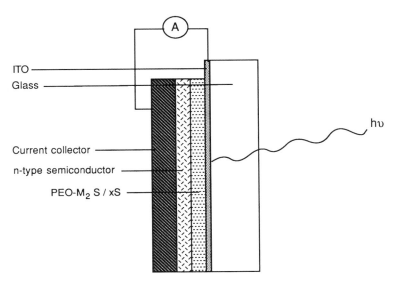

Figure 1.19. Schematic diagram of a photoelectrochemical cell utilizing a polymer electrolyte, as described in the text.

potassium and decreasing with the polysulfide concentration in the polymer. A close relationship between the conversion efficiency of the cell and the total ionic conductivity of the polymer electrolyte was also shown. Excellent stability under sustained white light illumination was demonstrated and was not a function of the cell performance, unlike the case of cells using aqueous solutions. An intercalation oxide $Na_x VO_y$ was used as a reference electrode for these cells.[85] The material is chemically stable as a function of time and temperature in the solid-state sodium–polyether electrolyte containing polysulfide in the dark and under continuous illumination. The stable reference electrode allowed characterization of the behavior of both CdSe and ITO electrodes, which permitted the determination of Mott–Schottky plots, yielding the semiconductor flatband potential and carrier concentration.

Energy conversion efficiencies observed so far are low, approximately 1 to 2%, but the conversion appears to be extremely stable after a steady regime is attained. The technology is still in its early stages, however, and with refinements in electrolyte architecture, will doubtlessly come the improved performances.

1.7. Sensors

All-solid-state designs for ionic sensors are favorable for mass production of miniaturized devices but are also important if, for example, high-temperature operation is required. Armand and co-workers[87] have proposed a PEO-based polymer electrolyte as a substitute for the internal liquid ionic phase in an electrochemical sensor, such as shown schematically in Figure 1.20. The electrochemical continuity between the metal and the ionic membrane is provided by an ionic solution. Various solid-state structures such as $Ag/AgF/LaF_3$ have been proposed as an ionic bridge[88]

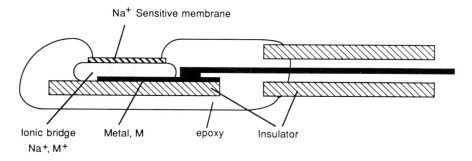

Figure 1.20. Schematic diagram of a solid-state ionic sensor. From Ref. 87.

but properties such as plasticity, allowing the maintainence of good contact, give polymer electrolytes a clear advantage for this purpose. For this particular study, NASICON ($Na_3Zr_2Si_2PO_{12}$) was used as the sensitive membrane and PEO was doped with sodium ions and the iodides of either copper or silver and compared with earlier prototypes using AgI–NaI or AgCl–NaCl mixtures as the ionic bridge.[89] Response times were found to be very short, similar to those of AgI–NaI cells and better than those of AgCl–NaCl cells. Voltage stability of the silver system (stable for 300 h) was better than that of the copper one where a drift value of approximately 0.2 mV h^{-1} was found. This value, however, was better than the 2 mV h^{-1} reported for the AgI–NaI system.

1.8. Conclusion

The highly successful application of polymer electrolytes to lithium rechargeable batteries has brought out the qualities of these materials in surface devices. These applications require the electrolyte to be stable under severe reducing, oxidizing, and basic conditions and may also require high operational temperatures. The chemical stability of the ether functional group strengthens the superiority of PEO as a host for a SPE. Although the conductivities at room temperature are poor, the limitations imposed on a polymer electrolyte do not relate only to its conducting properties; many functional groups, for example, alcohol, amino, keto, may not be tolerated under real working conditions. Equally, potentially reactive anions would be of little use. Such restrictions narrow the field of polymer electrolytes applicable in battery technology, and although it will be seen in Chapter 6 that excellent conducting properties can be achieved for electrolytes formed from the many "designed" polymer hosts, the majority have not been studied with respect to their electrochemical stability and many ultimately will never fit the criteria for application purposes. Another limitation in cell performance is the presence of mixed anion and cation transport in these electrolytes. The poor understanding at the present time of the actual mode of ion transport makes it difficult to foresee yet how to optimize cation transport without having a detrimental effect on other critical parameters. If current research and development work continues to progress, we could soon see polymer-based batteries for microelectronics applications and in the longer term, among other functions, mass electricity storage and the advent of large-scale use of electric vehicles.

References

1. M. Gauthier, M. Armand, and D. Muller, in *Electroresponsive Molecular and Polymeric Systems*, Vol. 1 (T. A. Skotheim, Ed.), Dekker, New York (1988), p. 41.

2. B. C. Tofield, R. M. Dell, and J. Jensen, *AERE Harwell Report 11261* (1984).

3. P. Passiniemi and O. Inganäs, *Solid State Ionics* **34** (1989), 225.

4. B. Scrosati, in *Polymer Electrolyte Reviews—1* (J. R. MacCallum and C. A. Vincent, Eds.), Elsevier, London (1987), p. 315.

5. M. Gauthier, D. Fauteux, G. Vassort, A. Bélanger, M. Duval, P. Ricoux, J. M. Chabagno, D. Muller, P. Rigaud, M. B. Armand, and D. Deroo, *J. Power Sources* **14** (1985), 23.

6. M. Gauthier, D. Fauteux, G. Vassort, A. Bélanger, M. Duval, P. Ricoux, J. M. Chabagno, D. Muller, P. Rigaud, M. B. Armand, and D. Deroo, *J. Electrochem. Soc.* **132** (1985), 1333.

7. A. Hooper and J. M. North, *Solid State Ionics* **9/10** (1983), 1161.

8. M. Gauthier, A. Bélanger, B. Kapfer, G. Vassort, and M. Armand, in *Polymer Electrolyte Reviews-2* (J. R. MacCallum and C. A. Vincent, Eds.), Elsevier, London (1989), P. 285.

9. G. Schwab and M. T. Lee, U.S. Patent 4,792,508 (Dec. 20, 1988).

10. M. T. Lee, D. Shackle, and G. Schwab, U.S. Patent 4,830,939 (May 16, 1989).

11. H. F. Hope and S. F. Hope, U.S. Patent 4,576,883 (Mar. 18, 1986).

12. H. F. Hope and S. F. Hope, U.S. Patent 4,808,496 (Feb. 28, 1989).

13. J. S. Lundsgaard, U.S. Patent 4,748,582 (May 31, 1988).

14. K. W. Semkow and A. F. Sammells, *J. Electrochem. Soc.* **134** (1987), 767.

15. G. Nazri, D. M. MacArthur, and J. F. Ogara, *Chem. Mater.* **1** (1989), 370.

16. G. A. Nazri, D. M. MacArthur, and J. F. Ogara, *Polym. Prep.* **30,** No. 1 (1989), 430.

17. K. M. Abraham, M. Alamgir, and S. J. Perrotti, *J. Electrochem. Soc.* **135** (1988), 535.

18. M. Alamgir, R. K. Reynolds, and K. M. Abraham, in *Materials and Processes for Lithium Batteries* (K. M. Abraham and B. B. Owens, Eds.), Electrochem. Soc., Pennington, N.J., 89-4, (1989), p. 321.

19. K. A. Fix and A. F. Sammells, in *Materials and Processes for Lithium Batteries* (K. M. Abraham and B. B. Owens, Eds.), Electrochem. Soc., Pennington, N.J., 89-4 (1989), p. 347.

20. P. S. S. Prasad, M. Z. A. Munshi, B. B. Owens, and W. H. Smyrl, *Solid State Ionics* **40/41** (1990), 959.

21. M. Z. A. Munshi and B. B. Owens, *Solid State Ionics* **26** (1988), 41.

22. C. K. Chiang, Y. W. Park, A. J. Heeger, H. Shirakawa, E. J. Louis, and A. G. MacDiarmid, *J. Chem. Phys.* **69** (1978), 5098.

23. P. J. Nigrey, D. MacInnes, D. P. Nairns, A. G. MacDiarmid, and A. J. Heeger, *J. Electrochem. Soc.* **128** (1981), 1651.

24. K. Kaneto, M. Maxfield, D. P. Nairns, A. G. MacDiarmid, and A. J. Heeger, *J. Chem. Soc. Faraday Trans.* **78** (1982), 3417.

25. M. B. Armand, *J. Phys.* **44(C3)** (1983), 551.

26. T. A. Skotheim (Ed.), *Handbook of Conducting Polymers*, Vols. 1 and 2, Dekker, New York (1986).

27. C. K. Chiang, *Polym. Commun.* **22** (1981), 1454.

28. T. Nagatomo, H. Kakehana, C. Ichikawa, and O. Omoto, *Jpn. J. Appl. Phys.* **24** (1985), L397.

29. T. Nagatomo, C. Ichikawa, and O. Omoto, *J. Electrochem. Soc.* **134** (1987), 305.

30. B. Scrosati, in *Solid State Ionic Devices,* (B. V. R. Chowdari and S. Radhakrishna, Eds.), World Science, Singapore (1988), p. 127.

31. C. Arbizzani and M. Mastragostino, *Electrochim. Acta* **35** (1990), 251.

32. F. Bonino, M. Ottaviani, B. Scrosati, and G. Pistoia, *Proceedings, Meeting of the Electrochemical Society, San Diego, Calif.* (1986).

33. Y. Geronov, B. Puresheva, P. Zlatilova, and P. Novak, in *Second International Symposium on Polymer Electrolytes* (B. Scrosati, Ed.), Elsevier, London (1990), p. 411.

34. G. Vassort, M. Gauthier, P. E. Harvey, F. Brochu, and M. B. Armand, in *Proceedings Symposium on Lithium Batteries, Honolulu, 1987,* 88-1, (A. N. Dey, Ed.), Electrochem. Soc., Pennington, N.J. (1988).

35. M. Gauthier, *First International Symposium on Polymer Electrolytes (ISPE-1), St. Andrews, June 17–19* (1987).

36. J. P. Gabano, in *Proceedings Symposium on Lithium Batteries, Honolulu, 1987,* 88-1, (A. N. Dey, Ed.), Electrochem. Soc., Pennington, N.J. (1988).

37. S. B. Brummer, V. R. Koch, and R. D. Rauth, in *Materials for Advanced Batteries* (D. W. Murray, J. Broadhead, and B. C. H. Steele, Eds.), Plenum Press, New York (1980), p. 123.

38. A. Bélanger, M. Gauthier, M. Duval, B. Kapfer, C. Robitaille, M. Robitaille, Y. Giguère, and R. Bellemare, *First International Symposium on Polymer Electrolytes (ISPE-1), St. Andrews, June 17–19* (1987), Extended Abstracts, p. 8.

39. M. Armand, *Faraday Discuss. Chem. Soc.* **88** (1989), 65.

40. D. V. Ragone, *Proceedings, Society of Automotive Engineers Meeting, Detroit, MI, May* (1968).

41. A. Hooper, in *Materials and Processes for Lithium Batteries* (K. M. Abraham and B. B. Owens, Eds.), Electrochem. Soc., Pennington, N.J., 89-4 (1989).

42. M. Z. A. Munshi, R. Gopaliengar, and B. B. Owens, *Solid State Ionics* **27** (1988), 259.

43. R. M. Dell, A. Hooper, J. Jenson, T. L. Markin, and F. Rasmussen, *Advanced Batteries and Fuel Cells,* EUR 8660 EN (1983).

44. M. Z. A. Munshi and B. B. Owens, *Solid State Ionics* **38** (1990), 87.

45. M. Z. A. Munshi and B. B. Owens, *Solid State Ionics* **38** (1990), 103.

46. M. Z. A. Munshi and B. B. Owens, *Solid State Ionics* **38** (1990), 95.

47. M. M. Nicholson and F. A. Pizzarello, *J. Electrochem. Soc.* **126** (1979), 1490.

48. F. B. Kaufman, A. H. Schroeder, E. M. Engler, and V. V. Patel, *Appl. Phys. Lett.* **36** (1980), 42.

49. C. J. Schoot, J. J. Ponjee, H. T. van Dam, B. A. van Doorn, and P. T. Bulwijn, *Appl. Phys. Lett.* **23** (1973), 65.

50. S. K. Deb, *Appl. Opt. Suppl.* **3** (1969), 192.

51. B. W. Faughan, R. S. Crandal, and P. M. Heyman, *R. C. A. Rev.* **36** (1975), 177.

52. W. C. Dautremont-Smith, *Displays* **3** (1982), 3.

53. G. C. S. Collins and D. J. Schiffrin, *J. Electroanal. Chem.* **139** (1982), 335.

54. K. Kaneto, K. Yoshino, and Y. Inuishi, *Jpn. J. Appl. Phys.* **22** (1983), 412.

55. P. Judeinstein, J. Livage, A. Zarudiansky, and R. Rose, *Solid State Ionics* **28–30** (1988), 1722.

56. W. C. Dautremont-Smith, *Displays* **3** (1982), 67.

57. T. Oi, *Annu. Rev. Mater. Sci.* **16** (1986), 185.

58. M. Kitao and S. Yamada, in *Solid State Ionic Devices* (B. V. R. Chowdari and S. Radhakrishna, Eds.), World Science, Singapore (1988), p. 359.

59. B. W. Faughan and R. S. Crandal, Display devices, in *Topics in Applied Physics*, Vol. **39**, Springer-Verlag, New York (1980), p. 181.

60. T. Oi and K. Miyauchi, *Mater. Res. Bull.* **16** (1981), 1281.

61. A. M. Glass, K. Nassau, and T. J. Negran, *J. Appl. Phys.* **49** (1978), 4808.

62. T. E. Haas, R. B. Goldner, G. Seward, K. K. Wong, G. Foley, and R. Kabani, *Proc. SPIE* (1987), 823.

63. R. B. Goldner, T. E. Haas, G. Seward, K. K. Wong, P. Norton, G. Foley, G. Berera, G. Wei, S. Schulz, and R. Chapman, *Solid State Ionics* **28–30** (1988), 1715.

64. M. Miyamura, S. Tomura, A. Imai, and S. Inomata, *Solid State Ionics* **3/4** (1981), 149.

65. M. Green and K. S. Kang, *Thin Solid Films* **40** (1977), L19.

66. G. G. Barna, *J. Electron. Mater.* **8** (1979), 153.

67. R. B. Goldner, in *Solid State Ionic Devices* (B. V. R. Chowdari and S. Radhakrishna, Eds.), World Science, Singapore (1988), p. 379.

68. R. B. Goldner, *Solid State Ionic Devices* (B. V. R. Chowdari and S. Radhakrishna, Eds.), World Science, Singapore (1988), p. 351.

69. O. Bohnke, C. Bohnke, and S. Amal, *Mater. Sci. Eng.* **B3** (1989), 197.

70. N. Kobayashi, Ph.D. thesis, Waseda University (1988).

71. S. Passerini, B. Scrosati, A. Gorenstein, A. M. Andersson, and C. G. Granqvist, *J. Electrochem. Soc.* **136** (1989), 3394.

72. S. Pantaloni, S. Passerini, and B. Scrosati, *J. Electrochem. Soc.* **134** (1987), 753.

73. O. Bohnke, *Second International Symposium on Polymer Electrolytes (ISPE-2), Siena, June 14–16* (1989), Extended Abstracts, p. 144.

74. F. Defendini and M. Armand, *First International Symposium on Polymer Electrolytes (ISPE-1), St. Andrews, June 17–19* (1987).

75. F. Defendini, Ph.D. thesis, INP Grenoble (1987).

76. M. Armand, D. Deroo, and D. Pedone, in *Solid State Ionic Devices* (B. V. R. Chowdari and S. Radhakrishna, Eds.), World Science, Singapore (1988), p. 515.

77. W. Wixwat, J. R. Stevens, A. M. Andersson, and C. G. Granqvist, in *Second International Symposium on Polymer Electrolytes (ISPE-2)* (B. Scrosati, Ed.), Elsevier, London (1990), p. 461.

78. G. Betz and H. Tributh, *Prog. Solid State Chem.* **16** (1985), 195.

79. T. A. Skotheim and I. Lundstrom, *J. Electrochem. Soc.* **129** (1982), 894.

80. T. A. Skotheim and O. Inganäs, *Mol. Cryst. Liq. Cryst.* **121** (1985), 285.

81. T. A. Skotheim, M. I. Florit, A. Melo, and W. E. O'Grady, *Mol. Cryst. Liq. Cryst.* **121** (1985), 291.

82. O. Inganäs, T. A. Skotheim, and S. W. Feldberg, *Solid State Ionics* **18/19** (1986), 332.

83. A. F. Sammels and G. P. Ang, *J. Electrochem. Soc.* **131** (1984), 617.

84. A. K. Vijh and B. Marsan, *Bull. Electrochem.* **5** (1989), 456.

85. B. Marsan, D. Fauteux, and A. K. Vijh, *J. Electrochem. Soc.* **134** (1987), 2508.

86. B. Marsan, D. Fauteux, and A. K. Vijh, *Solid State Ionics* **28–30** (1988), 1058.

87. P. Fabry, C. Montero-Ocampo, and M. Armand, *Sensors Actuators* **15** (1988), 1.

88. T. A. Fjeldy and K. Nagy, *J. Electrochem. Soc.* **127** (1980), 1299.

89. P. Fabry, J. P. Gros, and M. Kleitz, *Symposium on Electrochemical Sensors, Rome, June 12–14* (1984).

CHAPTER 2

Homopolymer Hosts

Most of the early work carried out on polymer electrolytes and, indeed, the majority of reported studies to date concentrate on poly(ethylene oxide) (PEO) and related homopolymers as the host for a number of salt species. Although a wide variety of polymer structures have now been made in an attempt to optimize electrical properties and/or mechanical and electrochemical stability, many of these incorporate PEO as the coordinating component. The properties of homopolymers are highlighted here; "designed" polymer hosts are discussed in Chapter 6.

2.1. Polyethers

The series of polyethers $[—(CH_2)_mO—]_n$ show remarkable variations in physical properties with increasing number of methylene repeat units as can be seen from Table 2.1. These differences cannot be sufficiently explained by variations in chemical structure, that is, addition of one oxymethylene group; rather, the molecular conformation and crystal structure must be considered to play a very important role. Polyoxymethylene ($m=1$) is a tight helix containing nine chemical repeat units and five turns in the fiber period of 1.739 nm.[1] Carazzolo[2] proposed a 29/16 helix as a more accurate (though essentially the same) molecular model which was confirmed by Uchida and Tadokoro.[3] A skeletal model of the polymer is given in Figure 2.1. The crystal structures of the higher homologs, poly(trimethylene oxide) ($m=3$) and poly(tetramethylene oxide) ($m=4$), indicate that the chains adopt a planar zigzag conformation and not a helical structure.[5-7] Poly(ethylene oxide) ($m=2$), having a fiber repeat period of 0.712 nm and a *trans–trans–trans* conformation, can also form this planar zigzag conformation when the polymer chains are put under stress.[8]

2.1.1. Poly(ethylene oxide)

The crystal structure of PEO, shown in Figure 2.2, and based on X-ray diffraction data, was first proposed by Tadokoro et al.[10] The normal structure (compared with that in Figure 2.1 when the sample is under stress) is of a 7/2 helix, that is, seven ethylene oxide repeat units with two turns in the fiber period of 1.93 nm, as shown in Figure 2.1 along with the structure of polyoxymethylene. The PEO helix has a much more open structure than that of poly(oxymethylene). The other cell param-

Table 2.1. Data on Polyethers of the Type $[-(CH_2)_m-O-]_n$

m	Polymer	Formula	Melting Point (°C)	Density (g cm^{-3}) X-ray	Density (g cm^{-3}) Observed	Solubility	Hardness	Molecular Structure
1	Poly(methylene oxide)	$(-CH_2O-)_n$	180	1.506	1.40–1.45	Soluble m-chlorophenol at 89°C, hexafluoroacetone sesquihydrate at room temperature	Hard	9/5 and 2/1 helices
2	Poly(ethylene oxide)	$(-CH_2CH_2O-)_n$	66	1.234	1.20–1.22	Soluble H_2O, $CHCl_3$, CH_3CN at room temperature	Soft	7/2 helix
3	Poly(trimethylene oxide)	$(-CH_2CH_2CH_2O-)_n$	37	—	1.11	Soluble $(C_2H_5)_2O$ or C_6H_6 at room temperature Insoluble H_2O	Soft	Planar zigzag and two other types
4	Poly(tetramethylene oxide)	$(-CH_2CH_2CH_2CH_2O-)_n$	43	1.11	1.06	Soluble $(C_2H_5)_2O$ or C_6H_6 at room temperature Insoluble H_2O or alcohols	Soft	Planar zigzag

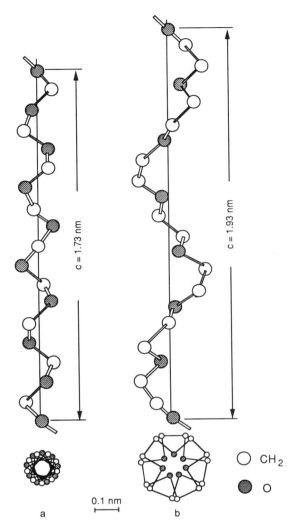

Figure 2.1. Schematic models of (**a**) polyoxymethylene and (**b**) poly(ethylene oxide). From Ref. 4, H. Tadokoro, *J. Polym. Sci. C*, Copyright (1966), John Wiley & Sons, Inc. Reprinted by permission of John Wiley & Sons, Inc., All Rights Reserved.

eters are reported to be $a = 0.805$ nm, $b = 1.304$ nm, and $\beta = 125.4°$, with the most probable space group $P2_{1/a}-C_{2h}$. Although the symmetry first proposed was isomorphous with the D_7 point group, a more detailed analysis showed that there was considerable distortion in the helical symmetry,[11] as a result of the influence of intermolecular forces. This and the ability of the polymer to orient when stressed[12] highlight the high degree of flexibility of the PEO chains. The following internal

b = 1.304 nm

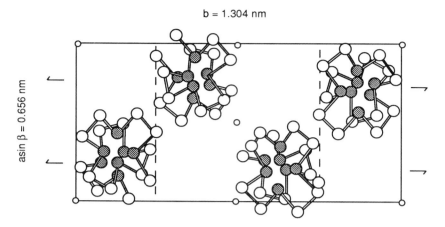

asin β = 0.656 nm

Figure 2.2. Crystal structure of PEO. From Ref. 9. Reprinted with permission from *Macromolecules* **6** (1973), 672. Copyright 1973 American Chemical Society.

rotation conformations for crystalline PEO have been assigned on the basis of extensive infrared and Raman studies[10,13–15]:

—O—CH$_2$—	*trans*
—CH$_2$—CH$_2$—	*gauche*
—CH$_2$—O—	*trans*

In the amorphous and molten states, the conformation is more disordered, with greater prevalence of an all-*trans* sequence. Below the melting point, high-molecular-weight PEO exists as a mixture of crystalline and amorphous phases. Numerous studies of the kinetics and mechanism of crystallization have shown this is dependent on polymer molecular weight, thermal history, and many other factors.[12]

The polymerization of ethylene oxide can be effected by two distinctly different processes. The first, known as oxyalkylation, depends on the tendency of ethylene oxide to oxyalkylate active-hydrogen sites in the presence of Lewis acid or base catalysts. The products are viscous liquids or waxy solids with a maximum molecular weight of approximately 10,000. These low-molecular-weight hydroxyl-terminated polymers are known as polyethylene glycols (PEGs). The second type of polymerization reaction involves the rapid polymerization of ethylene oxide to high-molecular-weight polymers on a catalytic surface in a heterogeneous reaction system. This was first accomplished by Staudinger and Lohmann[16] who prepared polymers of molecular weight 100,000 and higher by allowing the monomer to contact alkaline earth oxides for extended periods. Not until the late 1950s, with the discovery of much more active catalysts such as the alkaline earth carbonates,[17] did the commercial development of a series of high-molecular-weight PEOs occur. These Polyox resins (with molecular weights ranging up to 5,000,000) are produced by the Union Carbide Corporation. Other catalysts that have been reported in the literature include alkyl aluminum compounds,[18] hydrates of ferric chloride, bromide, and acetate,[19] and various alkyls and alkoxides of aluminum, zinc, magnesium, and calcium[12] along with various other inorganic salts.

Commercially available PEO, particularly the Polyox resin of molecular weight 4×10^6, is the material on which the vast majority of polymer electrolyte studies have been carried out to date. The polymer itself has a room temperature conductivity of the order of 10^{-9} S cm^{-1}. This relatively high conductivity may be the result of impurities, particularly Ca^{2+} and Mg^{2+} catalyst residues, which make it difficult to analyze with any certainty measurements on polymer electrolytes at very low salt concentrations. This point has been highlighted by Fauteux[20] with respect to the existence of a eutectic in the PEO–LiCF$_3$SO$_3$ system. It has been pointed out by Binks and Sharples[21] that it was found that after the passage of one coulomb of charge through a 250-mg sample of PEO, a current of several microamps still flowed, and it was suggested that, as the impurity content was unlikely to account for this, residual conductivity may result from an inherent proton generation process to provide charge carriers. Polymers containing oxygen bridges in the main chain, such as PEO and other polyethers, are susceptible to oxidative degradation. Hydroperoxide formation is accelerated by the presence of oxidizing agents, strong acids, polyvalent metallic ions such as Fe^{3+} and Cr^{3+}, and ultraviolet radiation.[12,22,23] Antioxidants are present in commercially prepared PEO that retard, but do not prevent, hydroperoxide formation with time. Other major impurities are amorphous silicon dioxide, which is added to assist the free flow of the polymer powder during production, and low-molecular-weight PEG fractions arising from the inhomogeneity of the molecular weight.

The melting point of PEO is dependent on the average molecular weight and molecular weight distribution of the sample. Values ranging from 60°C for molecular weight 4000 to 66°C for molecular weight 100,000 are commonly quoted. In general, the melting point increases with molecular weight up to 6000 and then levels off at a value of about 65°C.[22] The thermodynamic melting point of a perfect crystal of infinite dimensions has been calculated to be 76°C.[24]

The polymer unit cell dimensions lead to a calculated density of 1.33 at 20°C for PEO. Extrapolation of melt density data yields a density of 1.13 at 20°C for the amorphous phase. As the actual densities measured range between 1.15 and 1.26, the presence of numerous voids is implied. The melt density increases with molecular weight up to 20,000 and thereafter remains constant. This is probably due to the dilution of bulky end groups, which pack less efficiently, as the molecular weight increases.

The relationship between glass transition temperature (T_g) and molecular weight shows a rapid rise in T_g to a maximum value of -17°C for molecular weight 6000.[25,26] This maximum can be explained in terms of the percentage of crystalline material present in PEO, which is highest a molecular weight of about 6000. Beyond this point, chain entanglements lead to a lower degree of crystallinity. The T_g data for low-molecular-weight polymers parallel the density–molecular weight relationship when the polymers are quenched to the amorphous state.[25] For high-molecular-weight samples, the T_g value has been determined as between -65 and $-60°C$.

PEO exhibits compatibility with a wide range of plasticizers, low-molecular-weight compounds, and other polymeric materials.[27] At room temperature PEO is completely miscible with water in all proportions. Aqueous phases range from solutions containing less than 1% by weight of polymer, which have characteristic and useful rheological properties, to nontacky elastic gels at concentrations of around 20%, and finally to tough materials in which the water acts as a plasticizer. PEO is also soluble at room temperature in a number of common organic solvents such as acetonitrile, dichloromethane, carbon tetrachloride, tetrahydrofuran, and benzene.

2.1.2. Poly(propylene oxide)

After PEO, poly(propylene oxide) (PPO) has probably been the next most extensively used polymer in polymer electrolyte studies, largely because it is an amorphous material. Partially crystalline PPO may also be prepared by the polymerization of propylene oxide using diethyl zinc with water as cocatalyst.[28-30] Other catalysts include $FeCl_3$, water, and trimethyl aluminum.[31-33] As PPO has an asymmetric carbon atom, it can exist in isotactic form where successive asymmetric centers have the same configuration or in atactic form where the configurations are random. Depending on the polymerization conditions, the catalysts can act in a highly selective manner in ordering the configurations along the chain.

Crystal structures of isotactic PPO have been reported by Natta[34] and Stanley and Litt.[35] The chains were found to take up a zigzag conformation, bowing with an "S" shape in the direction of the a axis. This distortion may be caused by the proximity of a methyl group to a CH_2 group of the neighboring propylene unit on the chain, three atoms away. A planar zigzag structure would suffer from considerable strain. The methyl groups are 0.40 nm from the oxygen and 0.45 nm from the methylene groups of the neighboring chain. The orthorhombic unit cell as shown in Figure 2.3 has dimensions quoted as $a = 1.040$ nm, $b = 0.464$ nm, and $c = 0.692$ nm.

2.2. Poly(ethylene imine)

Poly(ethylene imine) (PEI), obtained by the cationic polymerization of ethylene imine or the cationic ring-opening polymerization of aziridene, is amorphous because the polymer is highly branched and crystallization is therefore suppressed.[36,37] The polymer chains contain a mixture of primary, secondary, and tertiary amino groups in the ratio of 1:2:1. Hydrolysis of poly(oxa-azoline)s yields linear PEI, which is a highly crystalline polymer, although the molecular weight of approximately 10^4 is relatively low.[38] Tanaka et al.[36] have described a synthetic route to high-molecular-weight (10^5) PEI that involves the acid-catalyzed debenzoylation of poly(N-benzoylethylamine), itself obtained by cationic ring-opening poly-

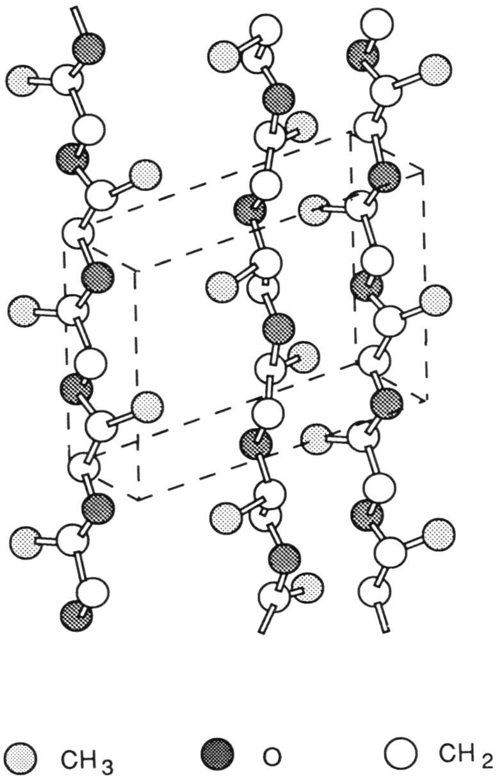

Figure 2.3. Structure of PPO polymer chains and their arrangement within the unit cell. From Ref. 35, E. Stanley and M. Litt, *J. Polym. Sci.*, Copyright (1956), John Wiley & Sons, Inc. Reprinted by permission of John Wiley & Sons, Inc., All Rights Reserved.

merization of 2-phenyl-2-oxazoline. This polymer had a sufficiently high molecular weight to allow a full crystal structure analysis to be carried out. Chatani et al.[39] revealed that linear PEI in the anhydrous state exists as double-stranded helices, each molecular chain adopting a 5/1 helical form (five monomeric units per turn), with an identity period of 0.958 nm. A pair of chains possessing the same helical sense form a double strand with a relative rotation of 180° about the chain axis at the same level. The formation of the double helix can be ascribed to N—H . . . N hydrogen bonding between the two chains. The crystals are orthorhombic, with cell dimensions $a = 2.98$ nm, $b = 1.72$ nm, and $c = 0.479$ nm, with eight strands in a unit cell. The structure is shown in Figure 2.4. The polymer is hygroscopic and forms at least two crystalline hydrates.[38,39] These adopt the fully extended form which reverts to the double-stranded form on dehydration. A high-molecular-weight equivalent of PPO, poly(*N*-methylaziridine), $(—CH_2CH_2N(CH_3)—)_n$, is extremely difficult to prepare because of competition between chain propagation and quaternization at the nitrogen atom.[40]

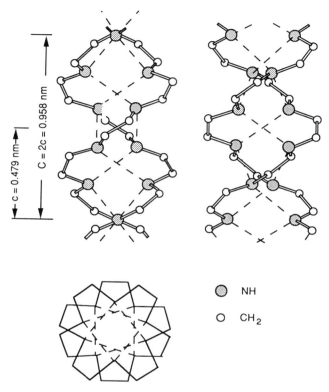

Figure 2.4. Double-stranded helical chains of PEI viewed along three perpendicular directions. Broken lines indicate N—H . . . N hydrogen bonds. From Ref. 39. Reprinted with permission from *Macromolecules* **15** (1982), 170. Copyright 1982 American Chemical Society.

2.3. Thia-alkanes

There have to date been few reports of the use of sulfur analogs of the polyethers in polymer electrolytes, possibly because many preparations result in insoluble crosslinked materials. Shriver described the preparation of polyalkylene sulfides, $(-CH_2CH_2)_mS$, where $m = 2–6$.[41] These were formed by reaction of the disodium salt of the appropriate dimercaptan with a dibromoalkane, following the method of Gotoh et al.[42] for the preparation of poly(pentamethylene sulfide). Molecular weights were reported as $m = 2$, 7900; $m = 3$, 6400; $m = 4$, 8600; $m = 5$, 10,000; and $m = 6$, 7000.

References

1. H. Tadokoro, T. Yasimoto, S. Murahashi, and I. Nitta, *J. Polym. Sci.* **44** (1960), 266.
2. G. A. Carazzolo, *J. Polym. Sci. A2* **1** (1963), 1573.
3. T. Uchida and H. Tadokoro, *J. Polym. Sci. A2* **5** (1967), 63.
4. H. Tadokoro, *J. Polym. Sci.* **15** (1966), 1.

5. K. Imada, T. Miyakawa, Y. Chatani, H. Tadokoro, and S. Murahashi, *Makromol. Chem.* **83** (1965), 113.

6. M. Cesari, G. Perego, and A. Mazzei, *Makromol. Chem.* **83** (1965), 196.

7. C. W. Bunn and D. R. Holmes, *Discuss. Faraday Soc.* **25** (1958), 95.

8. Y. Takahashi, Y. Osaki, and H. Tadokoro, *J. Polym. Sci. A2* **19** (1981), 1153.

9. Y. Takahashi and H. Tadokoro, *Macromolecules* **6** (1973), 672.

10. H. Tadokoro, Y. Chatani, T. Yoshihara, S. Tahara, and S. Murahashi, *Makromol. Chem.* **73** (1964), 109.

11. T. Takahashi and H. Tadokoro, *Makromol. Chem.* **6** (1973), 672.

12. F. E. Bailey and J. V. Koleske, *Poly(ethylene oxide),* Academic Press, New York (1976).

13. J. Maxfield and I. W. Shepherd, *Polymer* **16** (1975), 505.

14. J. L. Koenig and A. C. Angood, *J. Polym. Sci. A2* **8** (1970), 1787.

15. K. J. Liu and J. L. Parsons, *Macromolecules* **2** (1969), 529.

16. H. Staudinger and H. Lohmann, *Ann. Chim.* **505** (1933), 41.

17. F. N. Hill, F. E. Bailey, and J. T. Fitzpatrick, *Ind. Eng. Chem.* **50** (1958), 5.

18. R. A. Miller and C. C. Price, *J. Polym. Sci.* **34** (1959), 161.

19. M. E. Pruitt and J. M. Baggett, U.S. Patents 2,706,181 (1955) and 2,766,189 (1955).

20. D. Fauteux, in *Polymer Electrolyte Reviews—2* (J. R. MacCallum and C. A. Vincent, Eds.), Elsevier, London (1989), p. 121.

21. A. E. Binks and A. Sharples, *J. Polym. Sci. A2* **6** (1968), 407.

22. F. W. Stone and J. J. Stratta, *Encycloped. Polym. Sci. Technol.* **6** (1967), 103.

23. H. Vink, *Makromol. Chem.* **67** (1963), 105.

24. D. R. Beech and C. Booth, *Polym. Lett.* **8** (1970), 731.

25. B. E. Read, *Polymer* **1** (1966), 529.

26. J. A. Faucher, J. V. Koleske, E. R. Santee, J. J. Stratta, and C. W. Wilson, *J. Appl. Phys.* **37** (1966), 3962.

27. L. D. Berger and M. T. Ivison, in *Water Soluble Resins* (R. L. Davidson and M. Sittig, Eds.), Reinhold, New York (1962), p. 169.

28. J. Furukawa, T. Tsuruta, T. Sakata, T. Saegusa, and A. Kawaski, *Makromol. Chem.* **40** (1960), 64.

29. N. S. Chu and C. C. Price, *J. Polym. Sci. A* **1** (1963), 1105.

30. C. Booth, W. C. E. Higginson, and E. Powell, *Polymer* **5** (1964), 479.

31. C. C. Price and M. Osgan, *J. Am. Chem. Soc.* **78** (1956), 4787.

32. G. Gee, W. C. E. Higginson, and J. B. Jackson, *Polymer* **3** (1962), 231.

33. R. D. Colclough, G. Gee, and A. H. Jagger, *J. Polym. Sci.* **48** (1960), 273.

34. G. Natta, *Angew. Chem.* **68** (1956), 393.

35. E. Stanley and M. Litt, *J. Polym. Sci.* **43** (1956), 453.

36. R. Tanaka, I. Ueoka, Y. Takaki, K. Kataoka, and S. Saito, *Macromolecules* **16** (1983), 849.

37. G. E. Ham, in *Polymeric Amines and Ammonium Salts* (E. J. Goethals, Ed.), Pergamon Press, Elmsford, N.Y. (1980).

38. Y. Chatani, H. Tadokoro, T. Saegusa, and H. Ikeda, *Macromolecules* **14** (1981), 315.

39. Y. Chatani, T. Kobatake, H. Tadokoro, and R. Tanaka, *Macromolecules* **15** (1982), 170.

40. M. Armand, *Solid State Ionics* **9/10** (1983), 745.

41. S. Clancy, D. F. Shriver, and L. A. Ochrymowycz, *Macromolecules* **19** (1986), 606.

42. Y. Gotoh, H. Sakakihara, and H. Tadokoro, *Polym. J.* **4** (1973), 68.

CHAPTER 3

The Interaction Between Polymer and Salt

3.1. Ion Solvation by the Polymer

The existence of polar groups in polymers is a common feature; thus, it may be expected that they will behave as high-molecular-weight solvents and dissolve salts to form stable ion–polymer complexes. A salt dissolves in a solvent only if the associated energy and entropy changes produce an overall reduction in free energy of the system. This arises when the interaction between the ionic species and the coordinating groups on the polymer chain compensates for the loss of salt lattice energy.[1] The gain in entropy is brought about by destruction of the crystal lattice and also by gross deformations in the polymer structure. Some net decrease in entropy may arise form localized ordering of the polymer host by the ions. The enthalpy of solvation is essentially the result of electrostatic interactions between the cation positive charge and the negative charge on the dipolar groups of the polymer or of partial sharing of a lone pair of electrons on a coordinating atom in the polymer, leading to a coordinate bond. The lattice energy effects may be compensated for by such factors as a low value of cohesive energy density and vacancy formation, favored by a low glass transition temperature, Lewis acid–base interactions between the coordinating sites on the polymer and the ions, and long-range electrostatic forces such as cation–anion interaction energies. In addition, for a particular cation–polymer coordination group, the distance apart of the coordinating groups and the polymer's ability to adopt conformations that allow multiple inter- and intramolecular coordination are important. Neither poly(methylene oxide), $(CH_2O)_n$, nor poly(trimethylene oxide), $(CH_2CH_2CH_2O)_n$, for example, tend to form polymer electrolytes.[2] Both the rigidity of the chains in the former and the inability of either polymer to adopt low-energy conformations to maximize polymer–cation coordinations can account for this. Similarly, poly(propylene oxide) (PPO) is known to form less stable ion–polymer systems than poly(ethylene oxide) (PEO), because of the steric hindrance of the CH_3 group. In addition, intramolecular coordinations are more effective in poly(ethylene succinate), $(-OCH_2CH_2OCOCH_2CH_2CO-)_n$, which coordinates lithium ions more effectively than poly(ethylene sebacate), $-OCH_2CH_2OCO(CH_2)_8CO-$ where the coordinating groups are separated by eight methylene groups,[3] leading to greater dissociation. Table 3.1 summarizes the salt complexing abilities of PEO. It is

Table 3.1. Solubility of Salts in Coordinating Hosts, Principally PEO

	Li^+	Na^+	K^+	Rb^+	Cs^+	NH_4^+	Ag^+	Be^{2+}	Mg^{2+}	Ca^{2+}	Sr^{2+}	Ba^{2+}	Co^{2+}	Ni^{2+}	Cu^{2+}	Zn^{2+}	Hg^{2+}	Pb^{2+}
Cl^-	$+^a$	−	−	−	−	−	−		+	+	−	−	+	+	+	+		
Br^-	+	+	−	−	−	−	−		+	+	+	+	+	+	+	+	+	+
I^-	+	+	+	−	−	−	−	+	+	+	+	+	+	+	+	+	+	
SCN^-	+	+	+	+	+	+	+	+	+	+						+		
ClO_4^-	+	+	+	+	+	+	+		+	+	+	+	+	+	+	+		
$CF_3SO_3^-$	+	+	+	+	+	+	+		+	+	+	+	+	+	+	+	+	+
NO_3^-	+	+					+										+	+
BF_4^-	+	+																
BPh_4	+	+	+	+	+	+	+											
AsF_6^-	+	+									+							
PF_6^-	+																	
$H_2PO_4^-$	+																	

	$V^{2+/3+}$	Mn^{2+}	$Cr^{2+/3+}$	$Fe^{2+/3+}$	Al^{3+}	Ga^{3+}	Sc^{3+}	Y^{3+}	La^{3+}	Hf^{3+}	Ta^{3+}	W^{5+}	Ir^{3+}	Pt^{4+}	Th^{4+}	Zr^{4+}	Nb^{5+}	Mo^{5+}
Cl^-	+/+	+	+/+	+/+	+	+	+	+	+	+	+	+	+	+	+	+	+	+
Br^-	+/+	+	+/+	+	+		+	+	+		+							
I^-	+	+	+	+				+	+		+							
ClO_4^-	+	+		+	+													
$CF_3SO_3^-$	+	+	+	+	+				+									

	Cd^{2+}	In^{3+}	Sn^{2+}	Bi^{3+}	Ce^{3+}	Pr^{3+}	Nd^{3+}	Sm^{3+}	$Eu^{2+/3+}$	Gd^{3+}	Tb^{3+}	Dy^{3+}	Ho^{3+}	Er^{3+}	Tm^{3+}	Yb^{3+}	Lu^{3+}
Cl^-	+	+	+	+	+	+	+	+	+	+	+	+	+	+	+	+	+
Br^-	+		+	+	+	+	+	+	+	+	+	+	+	+	+	+	+
I^-	+		+		+	+	+	+	+	+	+	+	+	+	+	+	+
ClO_4^-	+		+			+	+									+	+
$CF_3SO_3^-$	+		+		+	+	+	+	+	+	+	+	+	+	+	+	+

^{a}+, Complex known or expected to be formed; −, no solubility.

apparent that PEO is able to solvate an extremely wide range of metal salts, including alkali, alkaline earth, lanthanide, and transition metals. The majority of these complexes are as yet either uncharacterized or only partially characterized.

Nitrogen heteroatom polyethers are likely to form stronger bonds with cations because of the higher donor number of the amine group (60 for N, 22 for O). Poly(ethylene imine) (PEI)–salt systems have been studied by Harris et al.[4] and Chiang et al. [5-7] The latter authors also studied the partially quaternized PEI and poly(N-acetylethylene imine), $[CH_2CH_2N(COCH_3)]_n$. Poly(N-methylaziridine) tends to form low conducting electrolytes, probably the result of extensive ion pairing, caused by the hindrance of the N-methyl group to complete cation coordination.

Silver salt complexes of poly(alkylene sulfide)s have been reported.[8,9] Electrolytes were formed by stirring the polymer in a methanol solution of the salt as the polymer itself was insoluble. Spacing between coordinating atoms appears less critical than for the oxides: electrolyte formation has been reported for poly(ethylene sulfide), poly(trimethylene sulfide), and poly(pentamethylene sulfide).[8,9]

3.2. Hard–Soft Acid–Base Principle

The hard–soft acid–base (HSAB) principle was suggested by Pearson.[10,11] as a means of accounting for and predicting the stability of complexes formed between Lewis acids and bases. Those acids and bases that are small, are highly electronegative, are of low polarizability, and are hard to oxidize are termed *hard,* that is, they tend to hold their electrons tightly; *soft* acids and bases are of low electronegativity, tend to be large, are highly polarizable, and are easy to oxidize, that is, they hold their valence electrons loosely. In general, a preference is found for complexes to be formed between hard acids and bases or soft acids and bases. In addition to classifying acids and bases according to the properties of the donor atom, it is also possible to group any given acid or base as hard or soft by its apparent preference for hard or soft reactants. For example, a base B may be categorized by the behavior of the following equilibrium:

$$BH^+ + CH_3Hg^+ \rightleftharpoons CH_3HgB^+ + H^+$$

In this competition between a hard acid (H^+) and a soft acid (CH_3Hg^+), a hard base will cause the reaction to go to the left but a soft base will cause it to proceed to the right. A listing of hard and soft acids and bases is given in Table 3.2. Molecules such as ethers and some amines that have donor atoms with high electronegativity and low polarizability are therefore hard bases. Consequently, PEO and PEI may be regarded as regular arrays of hard bases. Thioethers including poly(ethylene sulfide), on the other hand, are examples of soft bases.

The hardness and softness of acids and bases have been quantified through the softness parameter, σ_M. Table 3.3 lists these for cations and anions. The former is defined[12] as the difference between the ionization potential of the gaseous atom to form the cation and the enthalpy of hydration of the latter, normalized by subtraction of and division by the corresponding difference for the hydrogen ion. For the

Table 3.2. Classification of Hard and Soft Acids and Bases

Hard Acids	Hard Bases
H^+, Li^+, Na^+, K^+, (Rb^+, Cs^+)	NH_3, RNH_2, N_2H_4
Be^{2+}, $Be(CH_3)_2$, Mg^{2+}, Ca^{2+}, Sr^{2+}, (Ba^{2+})	H_2O, OH^-, O^{2-}, ROH, RO^-, R_2O
Sc^{3+}, La^{3+}, Ce^{4+}, Gd^{3+}, Lu^{3+}, Th^{4+}, U^{4+}	CH_3COO^-, CO_3^2, NO_3^-, PO_4^3
UO_2^{2+}, Pu^{4+}	ClO_4^-
Ti^{4+}, Zr^{4+}, Hf^{4+}, VO^{2+}, Cr^{3+}, Cr^{6+}, MoO^{3+}, WO^{4+},	F^-, (Cl^-)
$\quad Mn^{2+}$, Mn^{7+}, Fe^{3+}, Co^{3+}	
BF_3, BCl_3, $B(OR)_3$, Al^{3+}, $Al(CH_3)_3$, $AlCl_3$, AlH_3, Ga^{3+},	
$\quad In^{3+}$	
CO_2, RCO^+, NC^+, Si^{4+}, Sn^{4+}, CH_3Sn^{3+}, $(CH_3)_2Sn^{2+}$	
N^{3+}, RPO^{2+}, $ROPO^{2+}$, As^{3+}	
SO_3, RSO^{2+}, $ROSO^{2+}$	
Cl^{3+}, Cl^{7+}, I^{5+}, I^{7+}	
Hydrogen-bonded molecules	

Borderline Acids	Borderline Bases
Fe^{2+}, Co^{2+}, Ni^{2+}, Cu^{2+}, Zn^{2+}	C_6H_5, NH_2, C_5H_5N, N_3^-, N_2
Rh^{3+}, Ir^{3+}, Ru^{3+}, Os^{2+}	NO_2^-, SO_3^{2-}
$B(CH_3)_3$, GaH_3	
R_3C^+, $C_6H_5^+$, Sn^{2+}, Pb^{2+}	
NO^+, Sb^{3+}, Bi^{3+}, SO_2	

Soft Acids	Soft Bases
$Co(CN_5)^{3-}$, Pd^{2+}, Pt^{2+}, Pt^{4+}	H^-, R^-, C_2H_4, C_6H_6, CN^-, RNC, CO
Cu^+, Ag^+, Au^+, Cd^{2+}, Hg^+, Hg^{2+}, CH_3Hg^+	SCN^-, R_3P, $(RO)_3P$, R_3As
BH_3, $Ga(CH_3)_3$, $GaCl_3$, $GaBr_3$, GaI_3, Tl^+, $Tl(CH_3)_3$	R_2S, RSH, RS^-, $S_2O_3^{2-}$, I^-
CH_2, carbenes	
Pi acceptors	
HO^+, RO^+, RS^+, RSe^+, Te^{4+}, RTe^+	
Br_2, Br^+, I_2, I^+, ICN	
O, Cl, Br, I, N, $RO\cdot$, $RO_2\cdot$	
Metal atoms and bulk metals	

anion, the softness parameter is the difference between the electron affinity of the gaseous atom or radical forming the anion and the enthalpy of hydration of the latter, normalized by subtraction of the corresponding difference for the hydroxide ion and division by the difference between the ionization potential of the hydrogen atom and the enthalpy of hydration of the hydrogen ion. The data suggest that Mg^{2+} (hard), for example, would be expected to form very stable complexes with PEO, whereas Hg^{2+} (soft) would show only a weak interaction. Complexes of PEO with both these cations are readily formed but transference number measurements have demonstrated that Mg^{2+} ions are immobile[13] and Hg^{2+} ions are mobile[14] in PEO, which highlights a possible inverse-type relationship between promoting complex formation and the consequential effects on cation mobility.

For hard cations, including alkali and alkaline earth ions, the best electron donor groups follow the trend O>NR, NH>>S, which has been confirmed in stability constant measurements in cyclic polyether complexes.[15] For soft cations, those with partially filled *d* or *f* orbitals, there is a net preference for coordination by secondary

Table 3.3. Ionic Radii r_M, r_X and Softness Parameters σ_M, σ_X for Cations M^+ and Anions X^-

M^{z+}	r_M (nm)	σ_M	X^{z-}	r_X (nm)	σ_X
H^+		0.00	F^-	0.136	−0.71
Li^+	0.060	−0.95	Cl^-	0.181	−0.16
Na^+	0.095	−0.75	Br^-	0.195	0.10
K^+	0.133	−0.53	I^-	0.216	0.40
Rb^+	0.148	−0.49	OH^-	0.140	0.00
Cs^+	0.169	−0.46	SH^-	0.195	0.63
Cu^+	0.096	0.26	CN^-	0.182, 0.191	0.48
Ag^+	0.126	0.18	SCN^-	0.195, 0.213	0.84
Au^+		0.45	N_3^-	0.195	0.78
Tl^+	0.144	0.09	BF_4^-	0.232	
NH_4^+	0.148		NO_2^-	0.155	−0.24
Be^{2+}	0.031	−0.41	NO_3^-	0.189	−0.41
Mg^{2+}	0.065	−0.37	ClO_3^-	0.200	0.14
Ca^{2+}	0.099	−0.65	BrO_3^-	0.191	
Sr^{2+}	0.113	−0.59	IO_3^-	0.182	
Ba^{2+}	0.135	−0.60	ClO_4^-	0.236	0.00
Mn^{2+}	0.080	−0.11	MnO_4^-	0.229	
Fe^{2+}	0.075	−0.06	HCO_2^-	0.158	−0.44
Co^{2+}	0.072	−0.18	$CH_3CO_2^-$	0.159	−0.48
Ni^{2+}	0.070	−0.11	HCO_3^-	0.156	
Cu^{2+}	0.070	0.39	$B(C_6H_5)_4^-$	0.421	6.86
Zn^{2+}	0.074	0.37	S^{2-}	0.184	1.02
Cd^{2+}	0.097	0.59	Se^{2-}	0.209	
Hg^{2+}	0.110	1.28	CO_3^{2-}	0.185	−0.37
Sn^{2+}	0.093	0.31	SO_4^{2-}	0.230	−0.31
Pb^{2+}	0.132	0.58	SeO_4^{2-}	0.249	
Al^{3+}	0.050	−0.25	CrO_4^{2-}	0.256	
Sc^{3+}	0.075	−0.51			
Y^{3+}	0.093	−0.68			
La^{3+}	0.115	−0.65			
Gd^{3+}	0.094	−0.56			
Lu^{3+}	0.086	−0.67			
Pu^{3+}	0.101	−0.62			
Cr^{3+}	0.062	−0.06			
Fe^{3+}	0.060	0.22			
Ga^{3+}	0.062	0.29			
In^{3+}	0.081	0.44			
Tl^{3+}	0.095	0.92			
Bi^{3+}	0.102	0.61			
Ce^{4+}	0.080	−0.54			
Th^{4+}	0.099	−0.55			
U^{4+}	0.097	−0.38			

nitrogen groups (steric hinderance is usually encountered with tertiary nitrogen sites) giving a stability sequence of[16] NH>S>O. Poly(ethylene imine) has secondary amine groups. This polymer is a good complexing agent for transition metal ions as well as a solvating agent for alkali metal salts. Its solvating behavior is comparable to that of the cyclams.[17]

3.3. Anions

In polar solvents such as water or methanol, hydrogen bonding is important for specific anion solvation, whereas aprotic liquids and solvating polymers have negligible anion stabilization energies. Differences in the general solvation energies of anions do occur as the dielectric constant of the solvent varies. On passing from a polar, protic medium through to a less polar one, most anions are destabilized, the destabilized, the destabilization t ʒing greatest when the charge density and basicity of the ion[12] are low:

$$F^- >> Cl^- > Br^- > I^- \sim SCN^- > ClO_4^- \sim CF_3SO_3^- > BF_4^- \sim AsF_6^-.$$

The most suitable choices of anion for aprotic, low-dielectric-constant dipolar polymer-based polymer electrolytes are those to the right of the preceding series. These are large anions, with delocalized charge, are very weak bases, and possess low ion-dipole stabilization energies. In addition, their lattice energies are relatively low and they have little tendency to form tight ion pairs. These particular anions may be either soft (I^-) or hard (ClO_4^-) bases.

In general, the formation of polymer electrolytes is controlled by the cation solvation energy in opposition to the salt lattice energy. It is thus understandable that strongly solvated ions such as Li^+ can be complexed by PEO, even when the counterion is relatively small, like Cl^-, and there is an associated high lattice energy. The larger I^- anion is required for the heavier, less solvated K^+ ion. Divalent and trivalent cations again have sufficiently large solvation energies to induce the formation of complexes, even with the smaller Br^- anion. In addition, the salts formed from a soft cation and anion, for example, AgI, are not complexed by the ether oxygens, but a weaker hard–soft interaction allows for competition between the anion and the complexing polymer: $AgCF_3SO_3$ forms polyether-based polymer complexes but AgBr, and AgCl do not. The chloride, bromide, and iodide salts of Hg^{2+}, which is a much softer cation, all complex with ether oxygens; however, mercury halides are too covalent to allow the free Hg^{2+} ion to form in solution and therefore complexes are between polymer and molecular salt. Crystallographic studies of $PEO–HgCl_2$ complexes show the linear salt molecule to be slightly bent in the complex as a result of unsymmetrical interaction.[18,19] It is difficult to predict solely in terms of the HSAB principle which salts are likely to form polymer–ion complexes, but it does give a good indication as to the stability of the complex once it is formed.

Much useful information can be obtained on ion–polymer interactions from studies carried out on polyether–salt systems in solutions. Many of these were carried out prior to the initiation of studies on "polymer electrolytes" as such. Addition of salts to aqueous PEO solutions generally leads to a salting out of the polymer.[20,21] Ionic hydration is too strong for competitive ion–polymer interactions to be significant in this system. Conversely, Lundberg et al.[22] reported studies of the interactions of potassium salts with PEO and PPO in anhydrous methanol using dialysis and viscosity measurements. Methanol is a nonsolvent for PEO but addition of salts such as KI results in a salting-in effect. Other studies of methanolic solutions have been reported.[23-28] One study[28] investigated the binding of potassium and sodium thiocyanates and iodides to PEO in anhydrous methanol by fluorescence and ultrafiltration techniques along with conductivity measurements. It was shown conclusively that in addition to normal electrostatic forces, specific interactions led to a fraction of the anions binding to the PEO–cation unit. Arkhipovich et al.[29] investigated the solvation of Na^+, K^+, and Cs^+ by PEO in nitromethane which only weakly solvates cations. The equilibrium constants of the formation of polymer solvates were evaluated for different polymer molecular weights and were shown to increase with increasing chain length and decreasing cation radius. Estimates of the number of $-(CH_2CH_2O)-$ groups involved in the solvation shell was given as 6 to 12 depending on the cation. Many studies of PPO salt interactions have also been reported in the literature. Moacanin and Cuddihy[30] investigated the effects of adding up to 24.5% by weight of lithium perchlorate on the specific volume and viscoelastic behavior of low- and high-molecular-weight PPO. Wetton et al.[31,32] investigated the glass transition temperature of $PPO-ZnCl_2$ solutions and explained the rise in T_g with salt concentration as a result of intermolecular coordination of the zinc ions, resulting in crosslinking. Such studies in liquid solution have been valuable for shedding light on the nature of ion-solvent interactions in solvent-free polymer electrolytes.

3.4. Complex Formation

The term *complex* tends to be applied rather loosely when referring to polymer electrolytes. It is generally taken to mean the material formed when the polymer host interacts with the salt to form a new polymeric system. At high enough temperatures or in systems where crystallization is prevented, the ions are solvated by the polymer to form a homogeneous polymer–salt solution. With host materials such as high-molecular-weight linear PEO, however, the system crystallizes to form spherulites of well-defined stoichiometries. These "crystalline complexes" are often recognized by their melting points which can be well in excess of 100°C. The amorphous regions within the spherulites of complex material can be of a very different stoichiometry and it is therefore somewhat inappropriate to refer to the entire system as a "complex."

Wright[33] has used computational procedures to determine criteria for complex formation in PEO–sodium halide systems. Formation of PEO–NaX crystalline complexes involves the reaction of 1 mol of salt with 3 mol of ethylene oxide (EO) units. The lattice energy of the crystalline complex comprises several contributions:

$$E_{complex} = \epsilon(NaX) + \epsilon(3EO-NaX)_{intra} + \epsilon(3EO-NaX)_{inter} + \epsilon(3EO). \qquad (3.1)$$

$\epsilon(NaX)$ is the lattice energy of the salt within the complex, $\epsilon(3EO-NaX)_{intra}$ is the energy of interaction of the 3EO segment with ion pairs within the same polymer–salt molecular adduct, and $\epsilon(3EO-NaX)_{inter}$ is the energy of interaction of the segment with ion pairs of neighboring molecular adducts. $\epsilon(3EO)$ is the interaction of the segment with neighboring PEO helices in addition to changes in internal energy of this reference segment on complexation with the salt. A crystalline polymer–salt complex will thus form provided that

$$\epsilon(3EO-NaX)_{intra} + \epsilon(3EO-NaX)_{inter} + \epsilon(3EO) < E(NaX) + E(3EO)-\epsilon(NaX)$$

where $E(NaX)$ and $E(3EO)$ are the salt and polymer segment lattice energies, respectively. A value of -39.7 kJ (mol of 3EO unit)$^{-1}$ was calculated for the latter parameter.[34] Crystallographic data given by Chatani and Okamura[35] were used to compute the terms on the left-hand side of the preceding equation and the lattice energy of the ions within the complex. The computed results are plotted as functions of the partial electronic charges on the carbon atom, q_c, (q_o, the partial charge on the oxygen $= -2q_c$), in Figure 3.1. It can be seen that if a partial electronic charge of $+0.2$ to 0.23 on the carbon atom is assumed, then complex formation will occur as the total of the left-hand side terms is more negative than the right-hand side total. This value was taken from the charge distribution in the C—O bond employed by Mark and Flory[36] in their treatment of the dipole moments of PEO oligomers. Wipff et al.[37] considered that the oxygen atom has a partial charge of -0.6, that is,

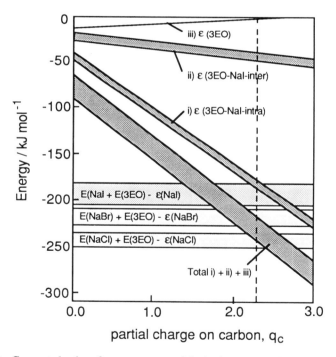

Figure 3.1. Computed values for components of the lattice energy of PEO–NaI as a function of the partial charge on carbon, q_c. The dark shaded area denotes the results for "high" and "low" van der Waals parameters. The "total" corresponds to the left-hand side of Eq. 3.1. The light shaded areas denote the range of results for the total of the terms on the right-hand side of Eq. 3.1 between "calculated" and "thermochemical cycle" values for E(NaI). Estimates for PEO–NaBr and PEO–NaCl are given in the lightest shading. From Ref. 34.

$q_c = 0.3$ in 18-crown-6 complexes with cations, thus allowing a greater stability margin for PEO–NaI. Estimates of corresponding energetic parameters for PEO–NaBr and PEO–NaCl also have plausible interpretations in terms of (as observed) complex formation in the former electrolyte and phase separation of the salt and polyether in the latter. A semiquantitative explanation for the solubility and insolubility of salts in polymers can thus be obtained in terms of lattice energies, although further calculations on other polymer–salt systems are obviously needed before a fully satisfactory theoretical prediction can be given.

References

1. C. A. Vincent, *Prog. Solid State Chem.* **17** (1987), 145.

2. M. B. Armand, J. M. Chabagno, and M. J. Duclot, in *Fast Ion Transport in Solids* (P. Vashista, J. N. Mundy, and G. K. Shenoy, Eds.), Elsevier North-Holland, Amsterdam (1979).

3. M. Watanabe, M. Rikuwa, K. Sanui, and N. Ogata, *Macromolecules* **19** (1986), 188.

4. C. S. Harris, D. F. Shriver, and M. A. Ratner, *Macromolecules* **19** (1986), 987.

5. T. Takahashi, G. T. Davis, C. K. Chiang, and C. A. Harding, *Solid State Ionics* **18/19** (1986), 321.

6. C. K. Chiang, G. T. Davis, C. A. Harding, and T. Takahashi, *Macromolecules* **18** (1985), 825.

7. C. K. Chiang, G. T. Davis, C. A. Harding, and T. Takahashi, *Solid State Ionics* **18/19** (1986), 300.

8. D. F. Shriver, B. L. Papke, M. A. Ratner, R. Dupon, T. Wong, and M. Brodwin, *Solid State Ionics* **5** (1981), 83.

9. S. Clancy, D. F. Shriver, and L. A. Ochrymowycz, *Macromolecules* **19** (1986), 606.

10. R. G. Pearson, *J. Am. Chem. Soc.* **85** (1963), 3533.

11. R. G. Pearson, *J. Chem. Ed.* **45** (1968), 581, 643.

12. Y. Marcus, *Ion Solvation,* Wiley, Chichester (1985).

13. L. L. Yang, A. R. McGhie, and G. Farrington, *J. Electrochem. Soc.* **133** (1986), 1380.

14. P. G. Bruce, F. Krok, and C. A. Vincent, *Solid State Ionics* **27** (1988), 81.

15. H. K. Frensdoff, *J. Am. Chem. Soc.* **93** (1971), 600.

16. J. R. Lotz, B. P. Block, and W. C. Fermelius, *J. Phys. Chem.* **63** (1959), 541.

17. M. R. Truter, in *Structure and Bonding.* No. 16. *Alkali Metal Complexes with Organic Ligands* (J. D. Dunitz, P. Hemmerich, C. K. Jørgensen, D. Reinen, J. A. Ibers, J. B. Neilands, and R. J. P. Williams, Eds.), Springer-Verlag, Berlin (1973), p. 71.

18. R. Iwamoto, Y. Saito, H. Ishihara, and H. Tadokoro, *J. Polym. Sci.* **6** (1968), 1509.

19. M. Yokoyama, M. Ishihara, R. Iwamoto, and H. Tadokoro, *Macromolecules* **2** (1969), 184.

20. F. E. Bailey and R. W. Callard, *J. Appl. Polym. Sci.* **1** (1959), 56.

21. F. E. Bailey and R. W. Callard, *J. Appl. Polym. Sci.* **1** (1959), 373.

22. R. D. Lundberg, F. E. Bailey, and R. W. Callard, *J. Polym. Sci. A1* **4** (1966), 1563.

23. K. J. Liu and J. E. Anderson, *Macromolecules* **2** (1969), 235.

24. K. J. Liu, *Macromolecules* **1** (1968), 308.

25. C. Detelier and P. Laszlo, *Helv. Chim. Acta* **59** (1976), 1333.

26. G. Chaput, G. Jeminet, and J. Juillard, *Can. J. Chem.* **53** (1975), 2240.

27. K. Ono, H. Konami, and K. Murakami, *J. Phys. Chem.* **83** (1979), 2665.

28. F. Quina, L. Sepuveda, R. Sartori, E. B. Abuin, C. G. Pino, and E. A. Lissi, *Macromolecules* **19** (1986), 994.

29. G. N. Arkhipovich, S. A. Dubrowskii, K. S. Kazanskii, N. V. Ptitsina, and A. N. Shupik, *Eur. Polym. J.* **18** (1982), 569.

30. J. Moacanin and E. F. Cuddihy, *J. Polym. Sci. Polym. Symp.* **14** (1966), 313.

31. R. E. Wetton, D. B. James, and W. Whiting, *J. Polym. Sci. Polym. Lett.* **14** (1976), 577.

32. D. B. James. R. E. Wetton, and D. S. Brown, *Polymer* **20** (1979), 187.

33. P. V. Wright, *Polymer* **30** (1989), 1179.

34. P. V. Wright, in *Polymer Electrolyte Reviews—2* (J. R. MacCallum and C. A. Vincent, Eds.), Elsevier, London (1989), p. 61.

35. Y. Chatani and S. Okamura, *Polymer* **28** (1985), 1815.

36. J. E. Mark and P. J. Flory, *J. Am. Chem. Soc.* **88** (1966), 3702.

37. G. Wipff, P. Weiner, and P. Kollman, *J. Am. Chem. Soc.* **104** (1982), 3249.

CHAPTER 4

Structure and Morphology

4.1. Crystalline Phases in Polymers

The vast majority of crystallizable polymers form spherulites as the dominant morphological entity during the transient stages of crystallization.[1] As their name suggests, these are spherically symmetric arrays of lamellar crystals immersed in the amorphous material. The lateral growth of polymer lamellae involves the apposition of fresh molecules in a regularly folded conformation along the lateral growth faces of crystals,[2] the long-chain molecules aligned more or less perpendicular to the planes of the lamellar fibrils. The thickness of the lamellae is governed by the average fold period in the constituent molecules. The apparently fibrillar structures in polymer spherulites are in fact due to highly elongated chain-folded lamellae, of the order of 10 to 100 nm in thickness, radiating from a central point, as depicted in Figure 4.1. These lamellae are not discrete crystals but many interlamellar linkages can occur when a crystallizing molecule nucleates on two different lamellae and may make a substantial contribution to the degree of crystallinity. Polymer chain axes in the spherulite are approximately perpendicular to the radius of the spherulite. A spherulite becomes a three-dimensional spherical object by virtue of the fact that lamellae occasionally branch. Spherulite radial growth continues until neighboring spherulites impinge on one another.

Depending on the nature of the system and experimental conditions, the layers of amorphous material between lamellae or at their boundaries either remain uncrystallized or crystallize very slowly to fill the structure. Thus the polymer is, to an approximation, a two-phase system (crystal plus supercooled liquid) over a wide range of temperatures. This behavior would be in contradiction to the phase rule if the phases were in thermodynamic equilibrium. This, however, is not the case: even at the "last" stages of crystallization, when the degree of crystallinity has reached its "limiting" value and the spherulites have long since ceased to grow, this secondary crystallization continues, never quite reaching completion.[2] Polymers can be successfully treated as two-phase systems in a state of torpid equilibrium, provided the time scale for measurements of interest is rapid with respect to any change in crystallinity.

The consequences of this for polymer electrolyte systems are important: it is extremely difficult for directly comparable materials to be prepared if crystalline phases are present. With numerous polymer electrolyte film preparation methods and subsequent thermal treatments reported in the literature, the variety of mor-

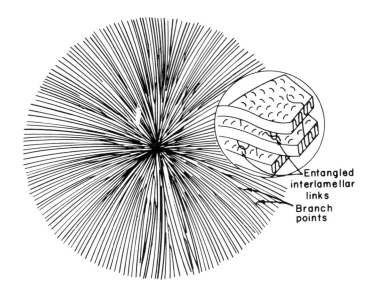

Figure 4.1. Schematic representation of a polymer spherulite with chain-folded lamellae. The polymer chain axes are more or less perpendicular to the radius of the spherulite. Noncrystallizable material accumulates between lamellae and at the outer boundary. From Ref. 1.

phological states formed must influence all the physical properties of the material and data reported.

Binary phase diagrams for numerous polymer–salt systems are described later. At the eutectic temperature, a reaction occurs to form a crystalline complex and a semicrystalline phase, SPEO, which is essentially poly(ethylene oxide) (PEO) with a variable quantity of salt dissolved up to a saturation value ether oxygen to sodium ion ratio (O:Na) equal to 42.5:1 for NaSCN in the intercrystalline amorphous regions.[3] For the purpose of constructing a phase diagram, this component may be treated as a single phase. This is acceptable, providing these interlamellar amorphous regions are reproducible. This appears to be so as, though SPEO formed in these systems is appreciably nonequilibrium in nature, the eutectic reaction for SPEO formation occurs under identical conditions, regardless of the salt concentration above the eutectic composition.

4.2. Preparation of Polymer Electrolyte Films

By far the most common method of producing polymer electrolyte thin films is by solvent casting. This involves preparing a solution of the polymer and salt: for PEO–salt systems, a 4% solution of the components in acetonitrile or methanol is most commonly used.[4,5] This solution is cast into formers on a polytetrafluoroethylene (PTFE) surface, the solvent is removed by slow evaporation, and the final film is heated under vacuum to remove residual solvent. Many polymer electrolytes are semicrystalline and, as discussed earlier, crystal formation may be significantly affected by the nature of the solvent, by the rate of solvent removal, and by traces of residual solvent which is capable of acting either as nucleation sites or as a plasticizer. In addition, the temperature at which films undergo final drying is important. Higher temperatures usually induce the formation of high-melting spherulites as the sample cools.[6] Consistency in casting techniques is important as demonstrated by the problems associated with phase diagram construction (Section 4.8), but in practice, no common criteria are universally observed.

To eliminate the effects of solvent, Gray et al.[7,8] developed a grinding/hot pressing technique for film preparation. This involves grinding the polymer to a fine powder under liquid nitrogen and subsequently milling appropriate quantities of the polymer and salt together to form an intimate mixture. CHN microanalysis has indicated that a homogeneous distribution can be achieved for O:cation concentration ratios as low as 50:1. After the samples were annealed at 120 to 150°C, they were cooled under pressure to form thin films. Conductivities were found to be significantly higher than those of solvent cast material, also annealed in this temperature region. A similar technique was applied by Lundberg et al.[9] who milled KI and PEO on a two-roll mill at 80°C. A good distribution was achieved after 10 to 15 min. Changes in melting point, T_g, and stiffness modulus of the material after various quantities of salt were milled into the polymer were interpreted as association of the KI with PEO.

Ito et al.[10] prepared 1-μm films of PEO–LiCF$_3$SO$_3$ by evaporating the components onto a silica substrate in a 1×10^{-4} Pa vacuum. The molecular weight of the polymer was initially 4×10^6, and after deposition, gel permeation chromatography indicated the molecular weight range to be 200 to 2000. The materials were waxlike and exhibited conductivities comparable to those of fully amorphous systems. Differential scanning calorimetry (DSC) traces showed diffuse endotherms in the temperature region -20 to 50°C.

Ultrathin polymer electrolyte films of the order of 1 μm in thickness have been prepared by plasma polymerization of tris(2-methoxyethoxy)vinyl silane.[11,12] A layer of plasma-polymerized monomer was first deposited on a stainless-steel, nickel, or gold substrate. This was sprayed with a 3% LiClO$_4$ solution in methanol followed by the deposition of a further polymer layer. The resulting layered structure was heated at 80°C for 24 h under vacuum to produce a homogeneous salt distribution. The plasma parameters had to be optimized to achieve maximum conductivity. The polymer contained Si—O—Si crosslinks and these increased in number as the radio frequency (RF) power to monomer flow rate and molecular weight ratio increased. These materials are discussed further in Chapter 6.

4.3. Solvent Deposited and Melt Recrystallized Films

Solvent-cast PEO-based films tend initially to have crystal lamellae 15 to 20 nm in thickness. Annealing high-molecular-weight PEO-based systems at temperatures close to the melting point of the particular crystalline phase leads to rapid lamellar thickening and to well-defined lamellae 40 to 60 nm in thickness. More complex "banded" structures may be observed, arising from the formation of stacks of lamellar fragments.[13]

Melt recrystallized phases have different morphologies from solvent cast materials, suggesting that there is a difference in fibril formation. For example, PEO–NaI films go through a series of morphological states on melt–recrystallization cycles, reducing lamellar thickness to about 20 nm. Extended annealing of this cannot bring about lamellar thickening which occurs rapidly in solution-cast films.[13] Chatani et al.[14] have observed at least three crystalline structures determined by X-ray crystallographic studies that depend on the method of sample preparation. It is also worth noting that the melting point of the polymeric system is dependent on crystal size, as well as crystal type, increasing with increasing annealing time.

4.4. Solvent Effects on Morphology

Wright and co-workers[13,15] showed that the solvent influenced the morphology of polymer electrolyte systems, producing material predominant in either high-melting, highly ordered (phase I) or low-melting, partially ordered (phase II) crystalline phases from the same stoichiometry. It was postulated that the predominating phase was controlled by competition between solvent and PEO for the salt. Low-polarity solvents would be unable to compete with PEO for the salt unlike solvents of high polarity, and thus highly ordered, high-melting PEO–salt complexes would be expected for the former systems. Neat[16] examined the effects of film-casting solvents of varying polarity on the structure and morphology of PEO–LiClO$_4$ and PEO–LiCF$_3$SO$_3$ systems. It was concluded that the morphologies were unaffected, in general, although tetrahydrofuran-cast films contained unique crystal phases. Type 2 spherulites in lithium trifluoromethanesulfonate-based films were converted from high- to low-melting crystals by being dissolved in dichloromethane. This effect may be the result of the volatility of the solvent, where rapid solvent removal favors low structural order.

It is difficult to generalize about effects of solvent type and rate of solvent removal as these are properties of the polymer–salt system. The trifluoromethanesulfonate and perchlorate materials serve as good examples of this point: slow solvent evaporation is essential in trifluoromethanesulfonate systems to resolve the spherulitic structures, as the crystallization kinetics are fast in comparison with the perchlorate system. Neat[16] reported that the morphology was independent of rate of solvent removal, probably because spherulite growth is not initiated until after solvent evaporation.

4.5. Trace Impurities

Low-molecular-weight impurities may have different and possibly conflicting effects on the system morphology. For example, they may act as plasticizers and increase the mobility of the polymer chains, giving rise to an enhancement in the conductivity. Because of the increased mobility in the chains, or because the solvent molecules may act as nucleation sites, the crystallization rate may increase and hence alter the distribution and ratio of crystalline to amorphous phases. If the proportion of the former increases, then the conductivity of the system would be expected to fall. In addition, if the solvent is highly polar, for example, water, it may bond strongly to the cation, interfering with the latter's interaction with and transference between the coordinating sites of the polymer. Rietman et al.[17] also noted that salts with oxygen-containing anions formed electrolytes whose conductivity could be improved by careful vacuum drying. This was ascribed to hydrogen bonding of the residual water to the oxoanions and the consequent reduction in their mobility.

Polyethers and the salts with which they complex to form polymer electrolytes are all hygroscopic to varying degrees. In addition to the problem of removing a casting solvent, it is very simple to introduce water as a solvent impurity into the polymer system. Acetonitrile and methanol, the two most commonly used casting solvents, are extremely difficult to purify to a completely "water-free" level. A number of studies have therefore been carried out to investigate the effects of water on the resulting morphologies and conductivities.

Weston and Steele[18] reported that small quantities of water in the casting solvent (acetonitrile) for PEO–LiClO$_4$ films gave rise to more amorphous electrolytes with a higher conductivity as compared with films cast under anhydrous conditions. Tanzella et al.[19] also noted an increase in conductivity on exposure of dry PEO–NaSCN, PEO–KSCN, and PEO–LiCF$_3$COO films to a laboratory atmosphere. Armstrong and Clarke[20] and Munshi and Owens,[21] however, found that the conductivity of PEO–LiCF$_3$SO$_3$ decreased with time when exposed to an atmosphere of strictly controlled low moisture levels. Hydration may restrict the mobility of the cation by forming a tight sheath around the ion, competing with the polymer chain coordinating sites and effectively reducing the number of cations participating in charge transport. Hydrogen bonding to the anion and the ether oxygens on the polymer backbone may also contribute to this effect.[17] Armstrong and Clarke noted, however, that a large increase in conductivity was observed if these films were exposed to laboratory air, whereas Munshi and Owens reported further decreases in conductivity when films were exposed to a saturated water vapor pressure. Multivalent cation-containing polymer electrolytes show greater behavioral dependence than monovalent systems when exposed to wet and dry conditions.[22] These are discussed in more detail in Chapter 7.

A study of water vapor absorption and conductivity in PEO and PEO–LiBF$_4$ films exposed to relatively high water vapor pressures (864–2806 Pa) was undertaken by Nicholson and Weissmuller.[23] A conductivity increase was observed only after the formation of the trihydrate salt. Data were interpreted as being consistent with the formation of a system of two microphases: a relatively stable nonconducting trihydrate phase and a conducting aqueous phase, saturated with both PEO and salt. Most of the water associated with this conducting phase could readily be removed from the film: this process was considered to be associated with a restructuring or recrystallization of the polymer.

4.6. Intercrystalline Amorphous Phases

Two amorphous phases have been recognized. One, an "intracrystalline" phase, is formed during crystallization of either the polymer–salt complex or pure PEO. This phase, which exists as microdomains embedded in or at the polymer–crystal boundary, results from chain entanglements (polymer chains in the amorphous phase may also form part of the crystalline one) and/or salt concentration inhomogeneities, preventing further crystal growth. Because ion transport and crystallization kinetics can be slow, this amorphous phase is not in thermodynamic equilibrium with the crystalline phase and salt concentration will vary significantly with crystallization mechanism. The "intracrystalline" amorphous phase is assumed to be dominant at low temperature, and ionic conductivity in polymer electrolytes below the eutectic temperature results from ion mobility through this phase. Hysteresis of conductivity measurements reported for many polymer electrolytes is generally the result of nonequilibration. The second amorphous phase, the "intercrystalline" phase, begins to form a significant proportion of the electrolyte only above the eutectic temperature and is in thermodynamic equilibrium with the crystalline complex in this instance. The salt concentration and proportion of amorphous to crystalline phase varies with temperature.

4.7. Polymer Electrolyte Structural Determination by EXAFS

Extended X-ray absorption fine-structure (EXAFS) spectroscopy provides useful short-range structural information, probing the local environment of a given element. Unlike Bragg diffraction techniques, information on the local environment in both crystalline and amorphous phases may be obtained. The EXAFS oscillations depend on the nature, distance from, and number of neighboring atoms and therefore it is simpler to study compounds containing heavy atoms because they are better backscatterers and give more pronounced interference patterns. The method adopted to analyze the EXAFS spectrum of an unknown structure is to produce a theoretical model, initially fitting the EXAFS spectrum of a known structure with a similar chemical environment to the unknown. Detailed information on the basic principles and data analysis of the EXAFS technique can be found in a number of books and reviews.[25–28]

Because the EXAFS technique has the ability to probe the local structure of both crystalline and amorphous material, it is clearly advantageous in the study of polymer electrolytes. The information obtained, however, relates only to the average local structure, and to obtain meaningful information, the system under study should be single phase. For some polymer electrolytes this is possible only at elevated temperatures, but for many systems fully amorphous complexes can be achieved at all temperatures.

To date, only a few EXAFS measurements on polymer electrolytes have been reported. These are the PEO–RbI, PEO–RbSCN,[29] PEO–CaI$_2$,[30] PEO–ZnI$_2$, and PEO–ZnBr$_2$[3,31] systems. Unfortunately, the results from each study are difficult to compare. Although the rubidium films were totally amorphous and dry, both the zinc and calcium halide systems were not rigorously dried by heating, to avoid the problem of high-melting spherulite formation.[32] The zinc-based films were prepared from totally anhydrous reagents, whereas the calcium-based electrolytes were prepared using the hydrated salt. For the latter material, oxygen nearest neighbors cannot be assigned in total to the polymer ether oxygens but may be partially ascribed to the cation's hydration sheath. Both types of polymer electrolyte film were found to be semicrystalline and consequently the EXAFS results are, in these instances, an average of at least two types of local environment experienced by the cations.

The studies of rubidium ion-containing polymers showed clearly that variations in the rubidium ion environment could be brought about by a change of anion. Experimental data for the PEO–RbX systems were least-squares fitted with four nearest oxygen atoms, eight carbon atoms in an outer shell, and an anion. For $X = $ I, the anion was calculated to lie at a radial distance of 0.38 nm from the cation, whereas the Rb—N distance was approximated as 0.30 nm for $X = $ SCN. For the iodide-containing material, the best fit gave two short Rb—O distances of 0.268 nm and two long Rb—O distances of 0.303 nm, and the carbon atoms were equally divided between two sites with Rb—C distances of 0.346 and 0.354 nm. With a thiocyanate anion, the best fit to experimental data gave only one Rb—O distance of 0.27 nm and two Rb—C distances of 0.324 and 0.355 nm. In these amorphous polymer electrolyte systems, the cations appear to occupy sites of fixed geometry, with temperature change having minimal effect. It was postulated that the nitrogen atom of the long SCN group protrudes into the first shell and may be responsible for disrupting the long and short Rb—O spacing observed in the iodide complexes. In addition, the Debye–Waller factors are considerably greater in the former case and there is a loss of spectral structure. A disruption of the chains would be in keeping with finding, as the larger parameter indicates, a system of relatively less order.

Table 4.1 give parameters for the local structure of zinc halide-based polymer electrolytes. For iodide complexes, a nearest-neighbor shell was proposed, containing four oxygen atoms and two iodide ions, whereas the best fit of experimental data for bromide complexes gave a structure consisting of six oxygens and two bromide ions.[31] Bromide K-edge EXAFS indicated little structure surrounding the anion, but it was deduced that each bromide ion had only one zinc cation nearest neighbor at a distance of 23.3 nm. For both rubidium and zinc halide systems, salt concentration (O:Rb = 4 or 8:1; O:Zn = 8–30:1) showed no influence on the local structure over the range studied.

The contribution of iodide ions in PEO–CaI$_2$ films to the backscattering and amplitude of the EXAFS oscillations was not found to be as great as that observed for PEO–RbI and PEO–ZnI$_2$ samples.[30] There was no evidence for the presence of iodide in the proposed first or second nearest-neighbor shells and the absence of calcium nearest neighbors in the iodide L$_{III}$ spectra substantiated this. This implies that there are no contact ion pairs in the P(EO)$_4$CaI$_2$ system in this particular study, unlike the PEO–RbI and PEO–ZnI$_2$ electrolytes. In this instance, however, the number of nearest-neighbor oxygen atoms was estimated as 10, implying that the

Table 4.1. Local Structure for PEO–ZnX$_2$ Electrolytes

Film	Casting Solvent	Nearest-Neighbor Atom to Z$_n$	Distance to Nearest Neighbor (nm)	Average Coordination Number	Debye–Waller Factor $\times$ 10^4 (nm^2)	Ref.
P(EO)$_6$ZnI$_2$	H$_2$O	O	0.2112	3.6	4.8	33
		I	0.2514	1.0	0.6	
P(EO)$_8$ZnI$_2$	MeCN/MeOH	O	0.2092	4.0	4.3	6
		I	0.2531	1.7	1.2	
P(EO)$_{10}$ZnI$_2$	MeCN/MeOH	O	0.2099	4.0	4.5	6
		I	0.2540	1.6	1.1	
P(EO)$_{12}$ZnI$_2$	MeCN/MeOH	O	0.2125	3.9	3.9	6
		I	0.2528	1.6	1.2	
P(EO)$_{15}$ZnI$_2$	MeCN/MeOH	O	0.2141	4.0	4.2	6
		I	0.2542	1.1	0.8	
P(EO)$_{20}$ZnI$_2$	MeCN/MeOH	O	0.2135	3.6	4.3	6
		I	0.2533	1.5	1.1	
P(EO)$_{30}$ZnI$_2$	MeCN/MeOH	O	0.2115	3.9	4.2	6
		I	0.2532	1.5	1.1	
P(EO)$_6$ZnBr$_2$	H$_2$O	O	0.2159	3.5	2.4	33
		Br	0.2324	2.4	1.0	
P(EO)$_{12}$ZnBr$_2$	MeCN/MeOH	O	0.2136	6.4	5.2	6
		Br	0.2339	2.1	1.2	
P(EO)$_{15}$ZnBr$_2$	MeCN/MeOH	O	0.2113	6.6	5.5	6
		Br	0.2339	1.8	1.1	
P(EO)$_{20}$ZnBr$_2$	MeCN/MeOH	O	0.2094	6.2	4.7	6
		Br	0.2333	2.2	1.4	
P(EO)$_{30}$ZnBr$_2$	MeCN/MeOH	O	0.2037	5.7	4.6	6
		Br	0.2338	1.8	0.9	
P(EO)$_{12}$ZnBr$_2$	MeCN/MeOH	Zn^a	0.2334	0.8	0.6	6
P(EO)$_{15}$ZnBr$_2$	MeCN/MeOH	Zn^a	0.2329	0.8	0.8	
P(EO)$_6$ZnCl$_2$	H$_2$O	O	0.2054	1.4	1.0	33
		Cl	0.2196	0.9	0.4	

^{a}Nearest-neighbor atom to Br.

cation lies in an extremely tight cage of ether oxygens or that some or all of the oxygens are due to a cation hydration sheath. This would also account for the absence of nearest-neighbor anions. It might be suggested, however, that a very different local structure would exist if rigorous drying procedures had been employed during the preparation of electrolytes.

A more detailed analysis of the EXAFS spectra of PEO–CaX$_2$ and PEO–ZnX$_2$ with data fitting for both oxygen and halide nearest neighbors has been reported by Latham et al.[33] Distances of the backscattering species from the target zinc species suggest that the fit corresponds to a "split" shell of two different nearest neighbors rather than a first- and second-shell nearest-neighbor situation. In addition, although it is possible to say with certainty that an interaction with oxygen occurs, it is more difficult to state a precise coordination number. Information on oxygen nearest neighbors is therefore qualitative rather than quantitative, and it still remains that in these materials calcium is surrounded by more oxygen nearest neighbors than zinc. Equally, it does not invalidate earlier conclusions that there is a modest oxygen coordination number dependence for calcium but not for zinc and that the oxygen coordination varies with the nature of the anion.

An interesting feature that arises from EXAFS studies of polymer electrolytes is that the structures determined in the crystalline and amorphous phases are effectively the same.[34] Although the crystalline phase also contains minor quantities of amorphous material, there is no indication of two superimposed spectra. Implications are that determination of crystal structures may give useful insight into the structural makeup of the ion-conducting amorphous phase. Further evidence for similarities in organization between the two phases has been suggested by Wright.[35] Entropies of fusion calculated from thermal analyses of $P(EO)_4NaI$ and $P(EO)_4NaSCN$ are approximately 2.5 J (mol EO)$^{-1}$ K^{-1}, no greater than for pure PEO crystals despite the additional salt component in the complex crystalline material. It was suggested that the complex melts were organized to some degree and/or the crystals were partially disorganized at their melting temperature. Evidence for a mesophase is discussed in Section 4.9, but evidence for the former in the case of PEO–NaSCN, at least, has been suggested by electron microscopy.[36]

4.8. Phase Diagrams

Interpretation of thermal and electrical behavior of linear PEO–salt systems has been greatly enhanced by characterization of polymer electrolyte crystalline and amorphous phases and the construction of phase diagrams. A general phase diagram for these systems was proposed by Sørensen and Jacobsen,[37] where it was assumed that a single stoichiometric crystalline phase such as $P(EO)_{3.5}LiCF_3SO_3$ existed along with amorphous polymer electrolyte (liquid), crystalline (to a first approximation) host polymer, and pure salt. Phase diagrams have been established for a number of polymer electrolytes through information derived from thermal measurements (DSC); optical microscopy, where polymeric crystal growth can be observed under a polarizing microscope; nuclear magnetic resonance (NMR), where the number of nuclei with a particular magnetic spin can be determined for each phase by determining the length of the T_2 decay time; and X-ray diffraction studies.

A knowledge of the microscopic structure and morphology of polymer electrolyte systems that form crystalline phases is essential, as they are only conducting to any great extent when in the elastomeric state. There is no straightforward relationship between conductivity and, for example, volume fraction of coexisting phases: there are too many other factors simultaneously involved, including nature, concentration, and mobility of charge carriers. Phase diagrams, however, offer a level of explanation for the dependence of conductivity on concentration and temperature.

A number of phase diagrams involving high-molecular-weight linear PEO have been constructed. These need to be interpreted with great care. Because of the sluggish nature of transport mechanisms and crystallization kinetics in polymers at low temperatures it is possible only to approach conditions of thermal equilibrium. In reality, there are always amorphous phases present that coexist with crystalline regions and can dissolve salt up to some saturation limit and which may not appear in the phase diagram. In addition, as discussed earlier, the thermal history affects the nature of the crystal spherulites and this in turn will be reflected in the phase diagram; however, the general behavior observed on cooling an amorphous polymer electrolyte can be readily understood by reference to these diagrams. The majority of phase diagrams share a number of general characteristics but discrete differences are observed that seem to resu't, at least in part, from the nature of the anion. In the following sections, the struc ure, morphology, and phase diagrams of polymer electrolytes are discussed with respect to particular anions.

4.8.1. Thiocyanates of Monovalent Cations

The problem associated with determining a complete crystal structure of the polymer electrolyte crystalline phases is that they generally have large unit cells with low symmetry. Parker et al.[38] studied fibers drawn from methanol solutions of 4:1 complexes of PEO and NaSCN and KSCN but could only observe a few low-angle reflections. The cell dimensions could be calculated and thus the number of ethylene oxide groups extending over the fiber repeat unit could be calculated. The unit cells contained eight EO units, two cations, and two anions. To account for this the authors suggested a double helix-type, two-chain unit cell, with each chain containing four units. It was noted, however, that the data could also be interpreted in terms of a simpler model in which the cations were equally coordinated on either side by two PEO chains.

Hibma[39] studied a range of crystalline complexes and was able to obtain much more detailed information by using Guinier photographs of the fibers oriented parallel to the focus line of the X-rays. From these he could determine the basal plane parameters and subsequently obtain the three-dimensional cell parameters by indexing a powder pattern of a randomly oriented film. A basal plane projection of a 4:1 PEO–KSCN structure is shown in Figure 4.2. The most important feature of this study is that the cations and anions are outside deformed polymer helices and are stabilized by interactions with ether oxygens on more than one chain. Chatani et al. are presently considering another model.[14]

The cell constants and space group of PEO–NaSCN have been determined by Chatani et al.[14] and confirm the data reported by Hibma.[39] The former group, however, reported polymorphism: at least three crystalline structures are observed and depend on the method of sample preparation. Crystallographic data are given in Table 4.2. Samples were prepared by either stretching films or immersing oriented PEO into an isopropanol solution of the salt. One structure (Form I) is shown in Figure 4.3. Unlike the potassium ions, the sodium ions are located along the axis of the helix. The sodium ions are accommodated in the helix at an interval of 0.360 nm, half of the fiber period. Each sodium ion is coordinated by four polyether oxygens and two anion nitrogens, a coordination number of 6. This differs from the 3:1 oxygen to sodium molar ratio of the crystalline complex of PEO–NaSCN determined by thermal studies,[39,40] although Lee and Crist[3] established the complex as $P(EO)_{3.5}NaSCN$.

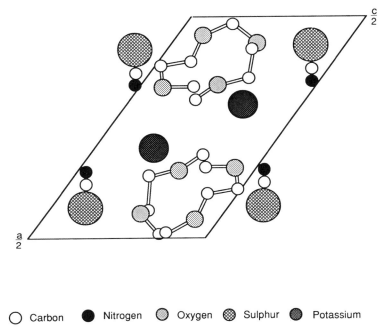

Figure 4.2. Model for the plane projection of the structure of P(EO)$_4$KSCN. A quarter of the unit cell is shown. From Ref. 39.

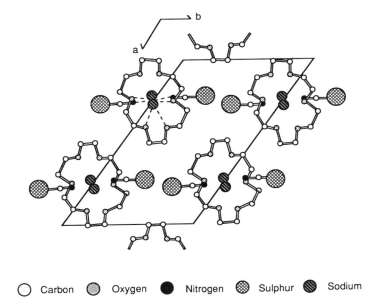

Figure 4.3. Crystal structure of PEO–NaSCN crystalline complex (Form I). From Ref. 14.

Table 4.2. Crystallographic Data for PEO–Salt Complexes

Electrolyte	a (nm)	b (nm)	c (nm) (Fiber Axis)	γ (deg)	ρ_{obs} (g cm^{-3})	Space Group	Ref.
PEO–NaSCN							
Form I	1.687	1.062	0.719	125.5	1.30	$P2_1/a$	14
Form II	1.550	1.302	0.583	100.0	—	$P2_1/a$	14
Form III	1.563	2.140	0.713	125.5	1.31	$P2_1/b$	14
PEO–KSCN	2.581	1.615	0.813	126.5	1.33	$B2/n$	14
PEO–NH$_4$SCN							
Form I	2.581	1.615	0.832	126.5	0.76	$B2/n$	14
Form II	0.831	1.199	0.633	103.0	—	$P2_1/a$ or $P2/a$	14
PEO–NaI	1.815	0.841	0.798	122.3	1.80	$P2_1/a$	45
PEO–NaBr	1.607	1.623	0.756	95.7	—	—	39
PEO–NaClO$_4$	1.727	1.702	0.840	115.0	1.41	—	39

Two phase diagrams have been proposed for the PEO–NaSCN system as a result of these studies.[3,40] Robitaille et al.[40] give the eutectic composition at 25:1, melting at 58°C, and Lee and Crist report it as 37.5:1, melting at 63°C. It has been tentatively postulated[40] that the low degree of crystallinity in the PEO-rich eutectic mixture reported by Lee and Crist may be due to entanglement effects and/or the result of high-temperature annealing of the samples. In one study,[3] no precautions were made to prevent contact with water vapor and this would also account for discrepancies between the two studies. For these reasons, and because of its completeness, the phase diagram given in Figure 4.4 is that attributed to Robitaille et al.

A similar phase diagram[40] is observed for KSCN systems and is shown in Figure 4.5. A number of different features exist. One peculiarity of the system is a crystallinity gap between the stoichiometries 8:1 and 12:1 and a partial crystallinity for all other compositional ratios greater than 4.5:1. Because of this, it was not possible to establish whether a eutectic mixture was formed with PEO; however, a 15:1 complex was found to exhibit an endothermic peak in the DSC trace at 56°C, below the PEO melting point, which suggests the possibility of a eutectic of similar composition to the NaSCN system. In contrast to the PEO–NaSCN system, all O:K stoichiometries studied, except a ratio of 15:1, could be cooled at a rate of 40°C min^{-1} and retain the amorphous phase. This can be explained by the slower diffusion rate of KSCN with respect to NaSCN, a consequence of the larger cation.

A high-temperature feature of a number of alkali metal salt–PEO systems that form crystalline compounds, including both the NaSCN and KSCN systems,[40–43] is a peritectic reaction that occurs above a particular temperature and composition in which the crystalline compound disproportionates into pure solid salt and a liquid phase termed the peritectic liquid. It appears that salt solubility is limited in the amorphous phase, the saturation salt concentration being less than in the crystalline complex. The disproportionation reaction takes place at 182°C for NaSCN complexes and at 95°C for KSCN complexes and corresponds to a peritectic equilibrium between solid P(EO)$_3$NaSCN, NaSCN, and a liquid phase and P(EO)$_{4.5}$KSCN,

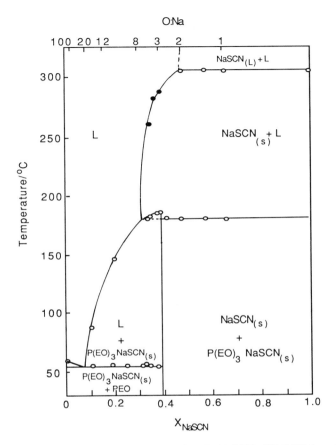

Figure 4.4. Temperature–composition diagram of PEO–NaSCN. (○) DSC data, (●) optical observations. From Ref. 40. Reprinted with permission from *Macromolecules* **20** (1987), 3023. Copyright 1987 American Chemical Society.

KSCN, and a liquid phase. The peritectic reaction for NaSCN complexes can be written as[40,41]

$$4(EO_3NaSCN)_s \rightleftharpoons 3(EO_4NaSCN)_l + NaSCN_s$$

The apparent coordination number of the KSCN complex in the liquid state is 5:1 and may be due to the larger size of the potassium ion. From phase diagram construction, above the peritectic equilibrium, salt solubility is described as "invariant." This temperature-independent behavior for salt solubility has been interpreted as an indication that well-defined solvates form in the amorphous PEO, resulting from a reaction similar in manner to crystalline compound formation and given by the equilibria

$$salt_s \rightleftharpoons salt_l$$
$$nEO_l + salt_l \rightleftharpoons (EO_nsalt)_l$$

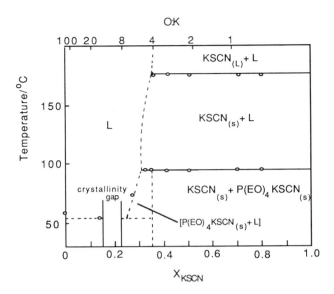

Figure 4.5. Temperature–composition diagram of PEO–KSCN. An additional tie line at 140°C corresponding to a crystalline transition for KSCN is not included. From Ref. 40. Reprinted with permission from *Macromolecules* **20** (1987), 3023. Copyright 1987 American Chemical Society.

where s and l denote solid and liquid phases. At disproportionation, an amorphous, saturated complex is formed and solid salt precipitates out. As the temperature is raised, redissolution is small over an approximately 50°C temperature range but shows a more "normal" dissolution curvature between 250 and 309°C, the melting temperature of the salt, which also coincides with phase separation of the salt from a liquid complex of composition EO:Na = 2:1. Had the salt not melted and phase separated, a more conventional curvature of the salt liquidus boundary may have resulted, where the region 182 to 250°C would have formed a final part of a curve from very much greater than 309°C downward. For both thiocyanate systems, the peritectic reaction is not reversible with both solid complex and solid salt precipitated on melt recrystallization. The metastable extension of the P(EO)$_3$NaSCN liquidus curve up to a composition of EO:Na = 3:1 ($x_{PEO} = 0.62$) may be explained in terms of a delayed disproportionation reaction resulting from the absence of solid NaSCN, present for EO:NaSCN > 3, which would provide nucleation sites for the precipitating salt. The dissolution–solvation scheme also applies to a PEO–CsSCN system.[41] Although no crystalline complex is formed, its solubility in amorphous PEO is also temperature independent over a significant temperature domain.

The crystal data for NH$_4$SCN, which is considered to be isomorphous with KSCN,[14] are given in Table 4.2. An approximate phase diagram for the PEO–NH$_4$SCN system is given in Figure 4.6. Thermal studies and X-ray diffraction results indicate that, as in the KSCN (and possibly the NaSCN) systems, the crystalline complex formed between PEO and NH$_4$SCN has a stoichiometry of 4:1

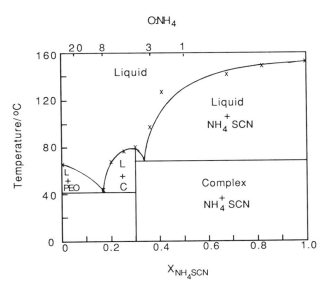

Figure 4.6. Approximate phase diagram for PEO–NH$_4$SCN. The complex P(EO)$_4$NH$_4$SCN at $x = 0.2$ may exist over a range of stoichiometries. From Ref. 44. Reprinted by permission of the publisher, The Electrochemical Society, Inc.

although the melting temperature of 68°C is much lower.[44] The eutectic composition is close to 8:1 and the temperature of the eutectic melt is 42°C. The crystallization kinetics for this system are very slow, of the order of weeks for 8:1 and 6:1 complexes and several days for a 4:1 system.

4.8.2. Halide Systems

A three-dimensional crystal structure of PEO–NaI has been determined by Chatani and Okamura.[45] The c and a projections of the structure are shown in Figure 4.7. The structure is similar to that of the PEO–NaSCN complex. The sodium ions are spaced at 0.399-nm intervals along the PEO helix and each sodium ion is coordinated, in this instance, by three ether oxygens and two iodide ions. Interatomic spacings are shown in Table 4.3. Crystallographic data are given in Table 4.2 along with data for the thiocyanate complexes. The proposed structure of the 3:1 complex indicates cation–anion spacing that is very close, implying strong interionic interactions, and that ionic motion in the crystalline phase is unlikely to occur. This agrees with other studies that have shown that polymer electrolyte conductivity is due to ionic motion in the more dilute amorphous phase.[46]

Partial phase diagrams have been reported for the PEO–NaI and PEO–LiI systems.[47–49] Both systems, shown in Figures 4.8 and 4.9, respectively, show the formation of a 3:1 intermediate complex. The PEO–NaI system behaves similarly to the PEO–NaSCN system in that this complex has a peritectic melting point at 190°C (182°C for the thiocyanate system). Both sodium salt systems, when crystallized

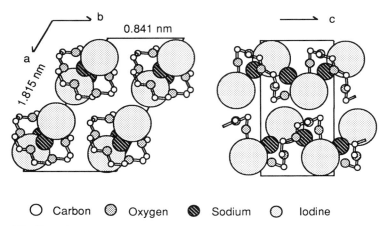

Figure 4.7. Crystal structure of P(EO)$_3$NaI. From Ref. 45. Reproduced by permission of the publishers, Heinemann Ltd. ©.

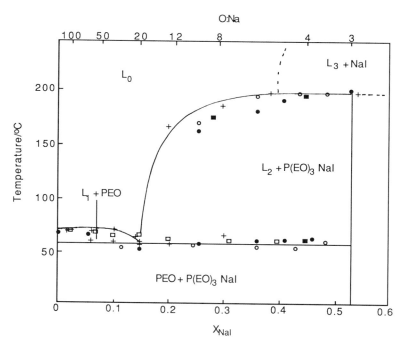

Figure 4.8. Phase diagram for PEO–NaI from DSC (■, ○, ●), optical microscopy (+), and conductivity (□) measurements. L$_{0-3}$ represent various amorphous phase compositions. Reprinted by permission of the publisher, The Electrochemical Society, Inc.

Table 4.3. Sodium–Oxygen and Sodium–Iodine Interatomic Distances

	Coordination Number	Distance (nm)
Na–O	4	0.222–0.259
	5	0.232–0.246
	6	0.225–0.278
	7	0.229–0.275
Na–O [P(EO)₃NaI]		0.224
		0.249
		0.261
Na–I		
Crystal	6	0.323
Gas		0.271
P(EO)₃NaI		0.295
		0.316

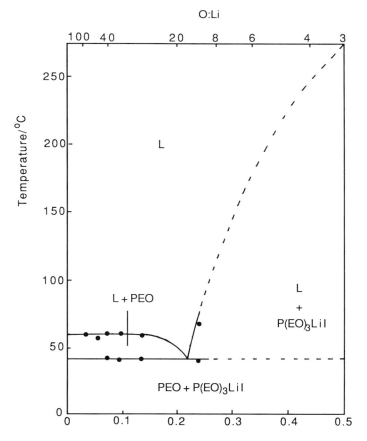

Figure 4.9. Phase diagram for PEO–LiI from optical microscopy measurements. From Ref. 49.

from the melt, form a eutectic near the composition $P(EO)_{20}NaX$, which melts at 55°C.

Yang et al.[50] have studied the complexes of PEO and KI by obtaining microscopic structural information from X-ray absorption near-edge structure (XANES) data. Spectral features of model compounds of known chemical structure, in which the K^+ ion was coordinated with 1, 2, and 6 oxygen atoms, were used as fingerprints for the polymer electrolytes with unknown coordination. Spectra of PEO–KI with a 4:1 stoichiometry were recorded over the temperature range 25 to 100°C. A change in structural features between 60 and 70°C was taken to be indicative of a change in coordination number. X-ray diffraction and thermal studies confirm that only crystalline complex exists for systems more concentrated than O:K = 8:1. DSC thermograms reveal that the complex melts over a wide temperature range, the onset being between 60 and 80°C. A change in the oxygen coordination therefore occurs as the complex melts and may indicate the occurrence of a peritectic reaction (incongruent melting), which was discussed earlier for the thiocyanate and NaI complexes.

4.8.3. Tetraphenyl Borate Systems

Besner et al.[41] have reported the phase diagram of the $PEO–NaB(Ph)_4$ system. As shown in Figure 4.10, it exhibits two consecutive peritectic equilibria, one at 138°C, in which a crystalline phase $P(EO)_8NaB(Ph)_4$ disproportionates into a new crystalline phase $P(EO)_5NaB(Ph)_4$, and a second at 165°C, in which the latter disproportionates to solid salt and a peritectic liquid of composition O:Na $\sim$ 8:1. This latter equilibrium is similar to that observed in PEO–NaSCN and PEO–NaI systems.

4.8.4. Halogen-Containing Complex Anionic Systems

The phase behavior of $PEO–LiAsF_6$ (Figure 4.11) has been studied by Robitaille and Fauteux[51] and Munshi and Owens.[52] It resembles that of $PEO–LiClO_4$ systems with crystalline phases of 6:1 and 3:1 stoichiometry. The melting points were found to be higher, however, with the 6:1 material melting at 136°C, some 70°C higher than the perchlorate system. Electrolytes with compositions above 5:1 were noted to be thermally unstable.[51]

Skeleton phase diagrams for PEO–LiX systems where $X = BF_4$, $AlCl_4$, and PF_6 have been established by Munshi and Owens.[52] Zahurak et al.[53] have also constructed a similar diagram for the $PEO–LiBF_4$ system, and the combined data are shown in Figure 4.12. A $PEO–LiBF_4$ electrolyte was found to be multiphase, with the stoichiometry of the first intermediate compound 2.5:1. This was deduced from the heats of fusion of excess crystalline PEO. A second crystalline complex had an O:Li ratio greater than 2.5:1, probably 4:1, the single intermediate crystalline phase reported by Payne and Wright.[15] The eutectic formed between PEO and $P(EO)_{2.5}LiBF_4$ occurs between the ratios 20:1 and 16:1.

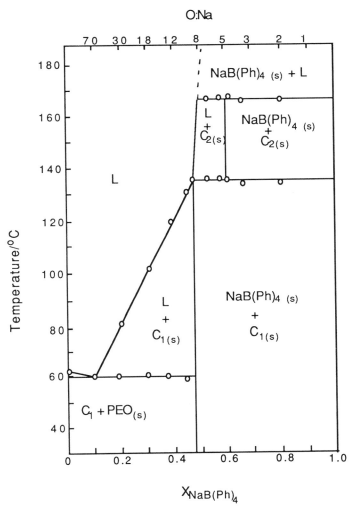

Figure 4.10. Phase diagram for PEO–NaB(Ph)$_4$. Crystalline complexes P(EO)$_8$NaB(Ph)$_4$ and P(EO)$_5$NaB(Ph)$_4$ are designated by C$_{1(s)}$ and C$_{2(s)}$, respectively. From Ref. 41.

PEO–LiAlCl$_4$ systems show unusual phase behavior. This may result from the fact that the salt decomposes without melting around 150°C. Uncomplexed PEO was detected throughout the composition range O:Li = 1:1 to 50:1 which may be the result of an appreciable solubility of the salt in PEO or, wholly or in part, to a partial dissociation to LiCl and AlCl$_3$.

The PEO–LiPF$_6$ system was reported to be isomorphous with PEO–LiAsF$_6$, up to an O:Li ratio of 6:1. Thereafter, the system becomes quite complex with multiphase formation. The first eutectic composition is given as 22:1 to 28:1, melting at 62.5°C. Second and third eutectic temperatures were reported to be 117 and 188°C, respectively.

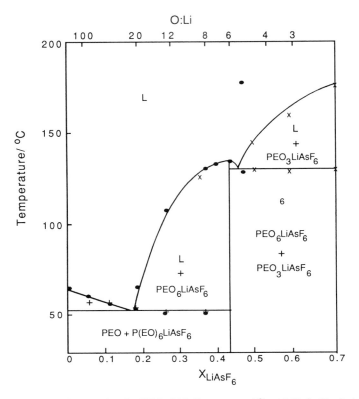

Figure 4.11. Phase diagram for the PEO–LiAsF$_6$ system. ($\bullet$, +) Ref. 51, ($\times$) Ref. 52. Reprinted from Ref. 51 by permission of the publisher, The Electrochemical Society, Inc., and from Ref. 52 by courtesy of Marcel Dekker, Inc.

4.8.5. Perchlorate Systems

Neat et al.[16,54,55] have studied spherulitic growth in PEO–LiClO$_4$ and PEO–LiCF$_3$SO$_3$ electrolytes by optical microscopy, scanning electron microscopy/energy dispersive X-ray analysis (SEM/EDX), and DSC. Three types of spherulites were identified:

Type 1 were fast growing, had a high salt content, and melted above 120°C.

Type 2 were slow growing, had a lower salt content, and melted between 45 and 60°C.

Type 3 were fast growing and were very similar to the host polymer, but had a lower melting point.

The growth rate of the spherulites is only a general statement as clearly the salt concentration of the surrounding amorphous material will influence the kinetics. For perchlorate-based systems, no more than two out of the three types of spherulite were present at any composition. At a composition of 4:1, large (250-μm) Type 1 spherulites were seen against a background of Type 2 structure. At 6:1, the Type 1

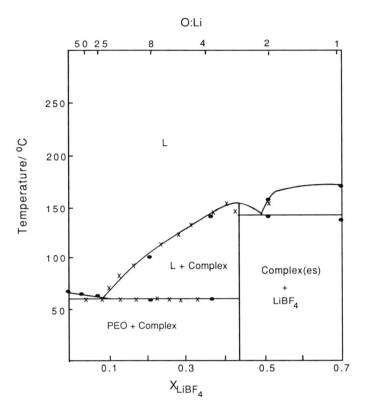

Figure 4.12. Phase diagram for the PEO–LiBF$_4$ system. (○) Ref. 52, (×) Ref. 53. Reprinted from Ref. 52 by courtesy of Marcel Dekker, Inc., and from Ref. 53 by permission from *Macromolecules* **21** (1988), 654. Copyright 1988 American Chemical Society.

spherulites had reduced to 50 μm in size with a concomitant increase in the proportion of Type 2 material. At composition ratios higher than 8:1, Type 1 spherulites were replaced by Type 3 structures. The size of these remained constant (~50 μm) but the number increased with the ratio.

X-ray diffraction results indicate that at least three distinct crystalline complexes are formed.[51] The proposed phase diagram for lithium perchlorate systems is given in Figure 4.13 and shows the existence of a 6:1 crystalline phase melting at 65°C and forming a eutectic with a melting point of 50°C and a composition around 10:1. No eutectic was observed between the 6:1 phase and the 3:1 phase, which melted at 160°C. It is possible that Type 1 spherulites are the 3:1 phase, Type 2 spherulites are a mixture of 3:1 and 6:1, and Type 3 may be identified with PEO. Similar phase behavior was reported by Ferloni et al.[56] although they suggested the eutectic melting point was some 20°C lower than that given by Robitaille.[51]

The thermal response of these systems is dependent on salt concentration and thermal history. Crystal phase formation is extremely slow in the salt concentration region with EO:Li ratio lower than 9:1 and was essentially the same as that of PEO

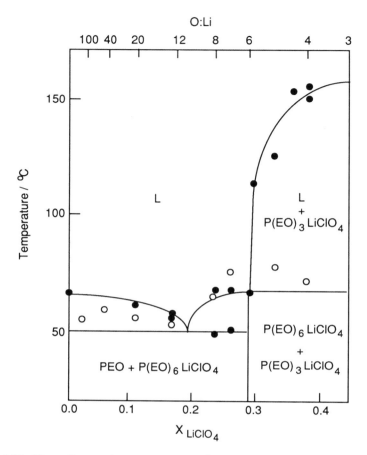

Figure 4.13. Phase diagram for PEO–LiClO$_4$. (●) Optical microscopy, (○) conductivity measurements. From Ref. 51. Reprinted by permission of the publisher, The Electrochemical Society, Inc.

for EO:Li ratios higher than 15:1, caused by the recrystallization of Type 3 spherulites.[16] Such behavior has been observed through conductivity studies[5,57] as well as other thermal investigations.[56]

4.8.6. Trifluoromethanesulfonate Systems

In general, it was found to be difficult to resolve the spherulitic structures in the triflate-based materials. This is due to the much faster crystallization kinetics than observed for the perchlorate-based systems and results in crystalline structures of low symmetry. This is true of many polymer electrolyte systems and is one of the problems that hinders the determination of complete structures of these crystalline phases; however, employing very slow solvent evaporation techniques[16] allowed full visual and EDX characterization. This provided evidence for the presence of all three spherulite types. A 4:1 complex was observed to be dominated by Type 1 spherulites. Between these regions, however, Type 2 and 3 spherulites were observed, the latter embedded in the former. A similar situation was observed for a 6:1 complex, confirmed by DSC studies that showed two high-temperature endotherms at 155 and 183°C. Distinction of the two melting points disappeared on thermal cycling, indicating that the kinetics of crystallization of Types 1 and 2 were not sufficiently high to retain the identity of each. Only when the molar ratio reached 3:1 were the Type 3 spherulites inhibited from growing.

The phase diagram for the PEO–LiCF$_3$SO$_3$ system was constructed by Robitaille and Fauteux from their own microscopy and conductivity data[51] and from thermal and NMR data of other workers[7,37,58-60] and has been extended to higher salt concentrations by Munshi and Owens.[52] This is presented in Figure 4.14. Zahurak et al.[53] have also studied this system down to a mole ratio of 1:1. The presence of a eutectic composition between 50:1 and 100:1 (100:1 to 200:1 is quoted by Munshi

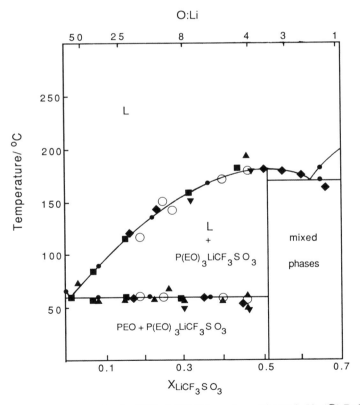

Figure 4.14. Phase diagram for the PEO–LiCF$_3$SO$_3$ system. (♦) Ref. 53, (○) Ref. 52, (▼) NMR, (○) DTA, DSC, (▲) conductivity, (■) optical microscopy data. From Ref. 51. Reprinted from Ref. 51 by permission of the publisher, The Electrochemical Society, Inc.; from Ref. 52 by courtesy of Marcel Dekker, Inc.; and from Ref. 53 by permission from *Macromolecules* **21** (1988), 654. Copyright 1988 American Chemical Society.

and Owens) and of one intermediate complex, $P(EO)_3LiCF_3SO_3$, was confirmed by optical microscopy observations for electrolytes of the appropriate compositions. The absence of a second complex at a 6:1 stoichiometry is likely due to the observation of a single average melting point, as described earlier; however, Zahurak reported a second phase of composition 7:1. This and the 3.5:1 phase coexist, the ratio of one to the other changing depending on the thermal history. The 3.5:1 phase was found to be the more stable. A further intermediate crystalline phase exists at salt concentrations higher than 3:1[52] and a second eutectic occurs around a stoichiometry of 2.2:1, melting at approximately 170°C. The uncertainty in the stoichiometry of the low-melting eutectic composition and, indeed, its actual existence has been highlighted by Fauteux.[61] High levels of impurities in commercially available PEO, of the order of the concentration of lithium trifluoromethanesulfonate at the eutectic point, may cause significant variations in apparent eutectic composition and depress the transition temperature. Work carried out on high-purity PEO still indicates the presence of a eutectic around a composition of 50:1.[62]

Stainer et al.[44] have reported thermal and X-ray diffraction data that support evidence for two crystalline complexes in the $PEO-NH_4CF_3SO_3$ system, corresponding roughly to a 4:1 complex and an 8:1 complex, melting at 41 and 60°C, respectively. At a 5:1 composition, both phases are present, but the low concentration phase is absent for a 3:1 composition. Interestingly, there is no evidence in the X-ray diffraction patterns for pure salt at 3:1 and 2:1 stoichiometries, but DSC traces of the high-salt-concentration systems suggest that a third, high-melting (94°C) phase may exist.

4.9. Mesogenic PEO–Salt Systems

Siddiqui and Wright[63] and Mussarat et al.[64] prepared a range of complexes with sodium and lithium phenolates, naphtholates, and sodium acrylate, some of which were shown to exhibit liquid crystalline properties. The PEO helical interval can vary between 0.36 and 0.42 nm to accommodate different anions and is therefore sufficient to accommodate the thickness of an aromatic ring in planar organic anions, which should stack alongside the helices. Complexes containing sodium (not lithium) salts with "hydrophobic" anions, that is, anions containing only hydrocarbon substituents such as phenolate, were found to deposit from methanol solution as thin films containing "macrodomains" of a uniaxially orientated structure up to approximately 0.5 cm in diameter. This lyotropic organization is probably the result of the mesogenic influence of the planar aromatic anions together with poor ion solvation, allowing the formation of precursor complexes with PEO prior to complete solvent removal. The formation of spherulites only with lithium salts of the organic acids may be the result of greater solvation. On heating above 60°C, reorganization to a microdomain (<1 μm) texture takes place, with final melting observed at approximately 140 to 200°C, depending on the anion and the stoichiometry. With melt recrystallization, spherulites form but there are no macro- or microdomains. With polar group-substituted anions, such as 4-chlorophenolate, spherulite morphologies have been observed on deposition from methanol solution.[64] The more solvated ions in this instance form complexes at later stages of solvent removal, preventing the large-scale organization necessary for macrodomain formation and reducing the stability of the precursor association with the polyether; however, thermotropic reorganization to microdomains occurs in these materials as well. The melting temperature is significantly reduced by introducing a polar substituent into the phenolate anion. Although the reason for this is not understood, it has been suggested by Wright and co-workers[64] that it may be the result of disorganization brought about by tautomerization and cation coordination on alternative sites of the anion.

The observation of mesogenic behavior in these complexes reflects related phenomena in the inorganic anion complexes.[64] Transition from macrodomain to microdomain structures takes place over the same temperature range as the onset of a mesophase in PEO–NaI and PEO–NaSCN[13] which was attributed to anion mobility. Prolonged annealing of PEO–NaSCN at submelting temperatures also gives rise to stacking of lamellar fragments. These latter systems, however, do not show the large-scale thermotropic reorganization observed in the organic complexes.

References

1. J. D. Hoffman, G. T. Davis, and J. I. Lauritzen, *Treatise on Solid State Chemistry*. Vol. 3. *Crystalline and Non-crystalline Solids* (N. B. Hannay, Ed.), Plenum Press, New York (1976), p. 497.

2. H. D. Keith and F. J. Padden, *J. Appl. Phys.* **34** (1963), 2409.

3. Y. L. Lee and B. Crist, *J. Appl. Phys.* **60** (1986), 2683.

4. J. E. Weston and B. C. H. Steele, *Solid State Ionics* **2** (1981), 347.

5. J. R. MacCallum, M. J. Smith, and C. A. Vincent, *Solid State Ionics* **11** (1984), 307.

6. M. Cole, M. H. Sheldon, M. D. Glasse, R. J. Latham, and R. G. Linford, *Appl. Phys. A* **49**, No. 3 (1989), 239.

7. F. M. Gray, J. R. MacCallum, and C. A. Vincent, *Solid State Ionics* **18/19** (1986), 252.

8. F. M. Gray, J. R. MacCallum, and C. A. Vincent, British Patent Application 8619049.

9. R. D. Lundberg, F. E. Bailey and R. W. Callard, *J. Polym. Sci. A1* **4** (1966), 1563.

10. Y. Ito, K. Syakushiro, M. Hiratani, K. Miyauchi, and T. Kudo, *Solid State Ionics* **18/19** (1986), 277.

11. Z. Ogumi, Y. Uchimoto, Z. Takehara, and F. R. Foulkes, *J. Electrochem. Soc.* **137** (1990), 29.

12. Z. Ogumi, Y. Uchimoto, Z. Takehara, and F. R. Foulkes, *J. Electrochem. Soc.* **137** (1990), 35.

13. C. C. Lee and P. V. Wright, *Polymer* **23** (1982), 681.

14. (a) Y. Chatani, S. Okamura, and Y. Fujii, *Polym. Prep.* **30**, No. 1 (1989), 404. (b) Y. Chatani, Y. Fujii, T. Takayanagi, and A. Honma, *Polymer* **31** (1990), 2238.

15. D. R. Payne and P. V. Wright, *Polymer* **23** (1982), 690.

16. R. J. Neat, Ph.D. thesis, Leicester Polytechnic (1988).

17. E. A. Rietman, M. L. Kaplan, and R. J. Cava, *Solid State Ionics* **17** (1985), 67.

18. J. E. Weston and B. C. H. Steele, *Solid State Ionics* **7** (1982), 81.

19. F. L. Tanzella, W. Bailey, D. Frydrych, G. C. Farrington, and H. S. Story, *Solid State Ionics* **5** (1981), 681.

20. R. D. Armstrong and M. D. Clarke, *Solid State Ionics* **11** (1984), 305.

21. M. Z. A. Munshi and B. B. Owens, *Appl. Phys. Commun.* **6** (1987), 299.

22. G. C. Farrington and R. G. Linford, *Polymer Electrolyte Reviews—2* (J. R. MacCallum and C. A. Vincent, Eds.), Elsevier, London (1989), p. 255.

23. M. M. Nicholson and T. P. Weissmuller, *J. Electrochem. Soc.* **132** (1985), 89.

24. P. A. Lee and J. B. Pendry, *Phys. Rev. B* **11** (1975), 2795.

25. B. K. Teo in *EXAFS: Basic Principles and Data Analysis* (C. K. Jørgensen, Ed.), Springer-Verlag, Berlin (1986).

26. E. Pantos and D. Firth, in *EXAFS and Near Edge Structure* (A. Bianconi, L. Incoccia, and S. Stipcich, Eds.), Springer-Verlag Berlin (1983).

27. J. Stöhr, in *Emission and Scattering Techniques* (P. Day, Ed.), Reidel, Dordrecht (1981).

28. A. V. Chadwick and M. R. Worboys, in *Polymer Electrolyte Reviews—1* (J. R. MacCallum and C. A. Vincent, Eds.), Elsevier, London (1987), p. 275.

29. C. R. A. Catlow, A. V. Chadwick, G. N. Greaves, L. M. Moroney, and M. R. Worboys, *Solid State Ionics* **9/10** (1983), 1107.

30. K. Andrews, M. Cole, R. J. Latham, R. G. Linford, H. M. Williams, and B. R. Dobson, *Solid State Ionics* **28–30** (1988), 929.

31. M. Cole, R. J. Latham, R. G. Linford, W. S. Schlindwein, and M. H. Sheldon, in *Solid State Ionics* (G. Nazri, R. A. Huggins, and D. F. Shriver, Eds.), **135** (1989), p. 383.

32. M. D. Glasse, R. G. Linford, and W. S. Schlindwein, *Br. Polym. J.,* in press.

33. R. J. Latham, R. G. Linford, and W. S. Schlindwein, *Faraday Discuss. Chem. Soc.* **88** (1989), 103.

34. R. G. Linford, *Faraday Discuss. Chem. Soc.* **88** (1989), 133.

35. P. V. Wright, *J. Macromol. Sci. A* **26** (1989), 519.

36. P. V. Wright, in *Polymer Electrolyte Reviews—2* (J. R. MacCallum and C. A. Vincent, Eds.), Elsevier, London (1989), p. 61.

37. P. R. Sørensen and T. Jacobsen, *Polym. Bull.* **9** (1982), 47.

38. J. M. Parker, P. V. Wright, and C. C. Lee, *Polymer* **22** (1981), 1307.

39. T. Hibma, *Solid State Ionics* **9/10** (1983), 1101.

40. C. Robitaille, S. Marques, D. Boils, and J. Prud'homme, *Macromolecules* **20** (1987), 3023.

41. S. Besner, A. Vallée, and J. Prud'homme, *Polym. Prepr.* **30,** No. 1 (1989), 406.

42. G. G. Cameron and M. D. Ingram, in *Polymer Electrolyte Reviews—2* (J. R. MacCallum and C. A. Vincent, Eds.), Elsevier, London (1989), p. 157.

43. M. C. Wintersgill, J. J. Fontanella, S. G. Greenbaum, and K. J. Adamic, *Br. Polym. J.* **20** (1988), 195.

44. M. Stainer, L. C. Hardy, D. H. Whitmore, and D. F. Shriver, *J. Electrochem. Soc.* **131** (1984), 784.

45. Y. Chatani and S. Okamura, *Polymer* **28** (1985), 1815.

46. C. Berthier, W. Gorecki, M. Minier, M. B. Armand, J. M. Chabagno, and P. Rigaud, *Solid State Ionics* **11** (1983), 91.

47. D. Fauteux, M. D. Lupien, and C. D. Robitaille, *J. Electrochem. Soc.* **134** (1987), 2761.

48. W. Gorecki, Ph.D. thesis, University of Grenoble (1987).

49. D. Fauteux, Ph.D. thesis, INRS (1986).

50. X. Q. Yang, J. Chen, C. S. Harris, T. A. Skotheim, M. L. Den Boer, H. Mei, Y. Okamoto, and J. Kirkland, *Mol. Cryst. Liq. Cryst.* **160** (1988), 89.

51. C. D. Robitaille and D. Fauteux, *J. Electrochem. Soc.* **133** (1986), 315.

52. M. Z. A. Munshi and B. B. Owens, *Appl. Phys. Commun.* **6** (1987), 279.

53. S. M. Zahurak, M. L. Kaplan, E. A. Rietman, D. W. Murphy, and R. J. Cava, *Macromolecules* **21** (1988), 654.

54. R. J. Neat, A. Hooper, M. D. Glasse, and R. Linford, in *Proceedings of the 6th Risø International Symposium on Metallurgy and Materials Science* (F. W. Poulsen, N. Hessel Andersen, K. Clausen, S. Skaarup, and O. T. Sørensen, Eds.), Risø National Lab., Roskilde (1984).

55. R. Neat, M. Glasse, R. Linford, and A. Hooper, *Solid State Ionics* **18/19** (1986), 1088.

56. P. Ferloni, G. Chiodelli, A. Magistris, and M. Sanesi, *Solid State Ionics* **18/19** (1986), 265.

57. A. Bouridah, F. Dalard, D. Deroo, and M. B. Armand, *Solid State Ionics* **18/19** (1986), 287.

58. B. L. Papke, M. A. Ratner, and D. F. Shriver, *J. Electrochem. Soc.* **129** (1982), 1434.

59. J. E. Weston and B. C. H. Steele, *Solid State Ionics* **2** (1981), 347.

60. M. Minier, C. Berthier, and W. Gorecki, *J. Phys.* **45** (1984), 739.

61. D. Fauteux, in *Polymer Electrolyte Reviews—2* (J. R. MacCallum and C. A. Vincent, Eds.), Elsevier, London (1989), p. 121.

62. D. Fauteux, J. Prud'homme, and P. E. Harvey, *Solid State Ionics* **28–30** (1988), 923.

63. J. A. Siddiqui and P. V. Wright, *Polym. Commun.* **28** (1987), 89.

64. B. Mussarat, K. Conheeney, J. A. Siddiqui, and P. V. Wright, *Br. Polym. J.* **20** (1988), 293.

CHAPTER 5

Aspects of Conductivity in Polymer Electrolytes

5.1. Total Direct-Current Conductivity

The total (dc) conductivity has routinely been used to characterize polymer electrolytes and a threshold value of approximately 10^{-5} S cm^{-1} has been used as the criterion for possible application purposes. A large number of materials reach such a conductivity value between room temperature and 100°C and some of these are shown in Figure 5.1.[1-11] Two aspects in particular govern the magnitude of the conductivity, the degree of crystallinity and the salt concentration, although the nature of the salt and polymer also has an important bearing.

5.1.1. Crystallinity in Polymer Electrolytes

The interpretation of the electrical conductivity of polymer electrolytes has been greatly complicated by the fact that many of the systems studied comprise more than one phase. Further difficulties have resulted from the action of slow phase equilibration processes. Particular phase distributions and compositions are highly dependent on factors such as method of preparation, impurities, and thermal history, as discussed in Chapter 4. Hysteresis in the conductivity is very common on thermal cycling because of slow crystallization kinetics, so that a downward temperature scan often reveals higher conductivities than the initial upward scan.[3,12,13] This can often be used to advantage where the time scale of the experiment is short and allows measurements to be made on the metastable system at temperatures well below where crystallization would normally complicate interpretation.[14,15] In addition to phase changes, transformations requiring the redistribution of salt between phases are common and often take place over a long time.

Berthier et al.[16,17] used nuclear magnetic resonance (NMR) techniques to assign nuclei to crystalline or elastomeric phases on the basis of the decay time of the transverse nuclear magnetization correlation function T_2. In Figure 5.2, the protons (polymer) and fluorines (salt) in the crystalline phase are shown as a function of temperature for $P(EO)_8LiCF_3SO_3$. The important conclusion that arose from these studies was that above the melting point of pure PEO, an equilibrium could be described between a stoichiometric crystalline phase and a more dilute amorphous

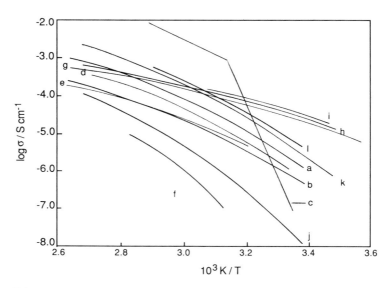

Figure 5.1. Temperature variation of the conductivity for (**a**) P(EO)$_8$LiClO$_4$,[1] (**b**) oxy-methylene-liked PEO$_{20}$LiClO$_4$,[2] (**c**) P(EO)$_{12}$NH$_4$CF$_3$SO$_3$,[3] (**d**) P(EO)$_8$LiClO$_4$ γ-ray crosslinked,[1] (**e**) P(PO)$_9$LiCF$_3$SO$_3$,[4] (**f**) poly(ethylene succinate)$_6$LiBF$_4$,[5] (**g**) poly(meth-yloligooxymethylene methacrylate)$_{18}$LiCF$_3$SO$_3$,[6] (**h**) (MEEP)$_4$LiCF$_3$SO$_3$,[7] (**i**) PEO–PPO–PEO block copolymer with LiCF$_3$SO$_3$ (8:1),[8] (**j**) P(EI)$_8$LiClO$_4$,[9] (**k**) poly(methyl cyanoethyl siloxane) with 10 wt% LiClO$_4$,[10] (**l**) ABA block copolymer with PEO grafted chains incorporating LiCF$_3$SO$_3$ (20:1).[11]

phase *that was responsible for the ionic conductivity*. As the temperature is raised, the crystalline phase progressively dissolves in the amorphous phase, thus increasing the concentration of charge carriers. Simultaneously, however, the polymer dynamics are significantly affected by the reduction in the amount of crystalline material and an increase in the transient crosslink density brought about by the increase in salt concentration in the conducting amorphous phase. The size and spatial distribution of nonconducting crystalline phases have a profound effect on the conductivity as they affect the tortuosity of the ion paths. In practice it is therefore extremely difficult to prepare directly comparable samples of polymer electrolytes unless they are fully amorphous systems.

Theoretical models for the conductivity of heterophase solid systems are generally based on effective medium theory, derived by Landauer,[18] or on the percolation theory of Kirkpatrick.[19] Considerable insight into the conductance behavior of heterogeneous polymer electrolytes can be gained by application of these theories using experimental conductivities and phase diagrams.[20] The concept of fractal geometry has been used by Le Mehaute[21] to discuss irreversible transfer of ionic carriers across electrode–electrolyte interfaces and the dynamic behavior of heterogeneous systems of electrochemical interest. Le Mehaute has applied fractal concepts to the effects of charge relaxation on ionic conductivity in polymer electrolytes and to the process of alterations in long-term storage.[21–23] Changes resulting from

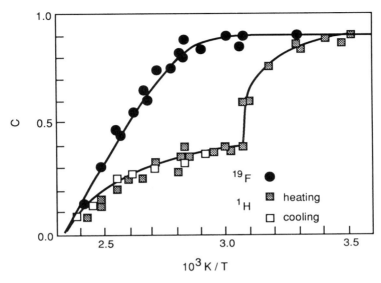

Figure 5.2. Fraction (C) of protons and fluorines belonging to the crystalline phase as determined from NMR (free precession) at various temperatures. From Ref. 16.

crystallization were described in terms of alterations in the fractal dimensions of the system; however, as it is generally accepted that significant ionic conductivity in polymer electrolytes is a property of amorphous phases above their glass transition temperatures (T_g), attempts to understand the basis of conductance behavior are being focused on studies of homogeneous amorphous materials. The effect of synthesizing polymers in which crystallization is impossible or suppressed is to markedly improve the transport properties at ambient temperatures where linear PEO is most crystalline (see Chapter 6); however, these modifications introduce a finite volume of material, mainly as crosslinks or backbone, that does not participate in conduction, and at temperatures above 100°C, where linear PEO-based systems are also amorphous, PEO may provide the best conducting matrix.

5.1.2. Salt Concentration

The ionic mobility is closely correlated to the relaxation modes of the polymer host. This can be observed through the increase in T_g of polymeric systems as salt concentration is increased. This reduction in segmental motion is usually interpreted as being a result of the effects of an increase in intramolecular and intermolecular coordinations between coordinating sites on the same or different polymer chains caused by the ions acting as transient crosslinks.[24] This is shown schematically in Figure 5.3. As well as conductivity reductions caused by the stiffening of the matrix, the availability of vacant coordinating sites is greatly reduced at high salt content. In addition, strong ion–ion interactions in systems of low permittivity such as the polyethers are probable and therefore ion migration is likely to involve the cooperative motion of several ions. The conductivity of a homogeneous polymer electrolyte phase may be given as

$$\sigma(T) = \sum_i n_i q_i \mu_i$$

where n_i is the number of charge carriers of type i, q_i is the charge on each, and μ_i is the mobility. This includes mobile charged aggregates. The relationship between charge carrier and salt concentration and the effects of ion association are discussed more fully in Chapter 9. Briefly, at low salt content, the mobility of ions is relatively unaffected by concentration, as the transient crosslink density will be low and therefore the conductivity will be controlled by the number of charge carriers. As the salt concentration increases, ion pairs and mobile higher aggregates are pre-

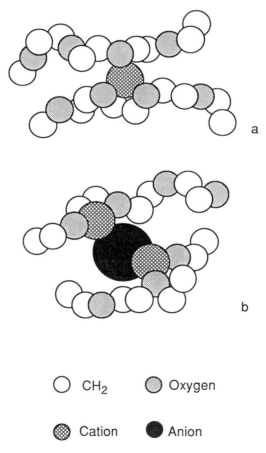

Figure 5.3. Formation of transient crosslinks via (**a**) cations and (**b**) a triple ion. The effect is to reduce segmental mobility and decrease the overall ion mobility.

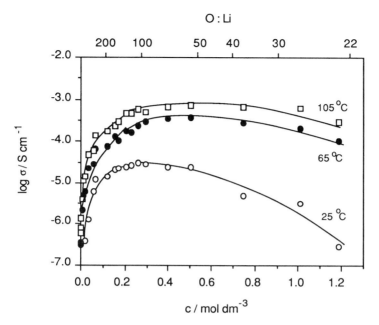

Figure 5.4. Variation in conductivity with salt concentration for an amorphous oxy-methylene-linked poly(ethylene oxide) polymer containing $LiClO_4$.

dicted to form,[25] which may then form higher, less mobile clusters and may also act as transient crosslinking species (Figure 5.3). At the highest salt concentrations, typically for an EO:salt molar ratio below 10:1, the system may be best thought of as a continuous "Coulomb fluid" where long-range interactions are important and where the material may be thought of as having much in common with a molten salt hydrate. Thus, the concentration dependence of the conductivity is a very complex function. In general, it has the shape illustrated in Figure 5.4, which here is for an amorphous oxymethylene-linked PEO–$LiClO_4$.[2]

5.2. Pressure Dependence of the Conductivity

A number of studies have been carried out where pressure rather than temperature has been used as a variable in investigations of the conductivity. Archer and Armstrong[26] found that the bulk conductivity of a $P(EO)_{4.5} LiCF_3SO_3$ electrolyte decreased with increasing pressure over the range 60 to 300 MPa. When a cell was held at a constant pressure and temperature, the conductivity was again seen to fall with time. These effects were attributed to some form of structural change at higher pressures and may possibly involve increased crystallization.[27]

Measurements of transport processes in crystalline solids as a function of pressure allow the activation volume V^* for the conduction process to be evaluated. It is defined by[28]

$$V^* = \left(\frac{\partial G}{\partial P} \right)_T \tag{5.1}$$

where ∂G represents the free energy change associated with the basic steps in the conduction mechanism. Values of V^* can then be used to test theoretical models of the process. This approach has had application in a wide number of solid systems and has been applied to polymer electrolytes by Fontanella and co-workers[29-34] and Chadwick et al.[35,36] Both isobaric and isothermal conductivity studies have been carried out. In determining the most appropriate method of association of the free energy change with the ionic conductivity, two options were discussed, one described by an Arrhenius expression, the other a free volume expression. With the assumption of Arrhenius behavior,

$$\sigma = \sigma_0 \exp(-E_a/RT) \tag{5.2}$$

where σ_0 is the preexponential factor and E_a is the activation energy, leads to an activation volume

$$V^* = -RT \left[\left(\frac{\partial \ln \sigma}{\partial P} \right)_T - \left(\frac{\partial \ln \sigma_0}{\partial P} \right)_T + \frac{1}{R} \left(\frac{\partial S}{\partial P} \right)_T \right] \tag{5.3}$$

where S is the entropy of activation. Assuming that entropy is independent of pressure and that pressure-dependent terms in σ_0 are the number of charge carriers and attempt frequency results in a complex expression which was approximated to

$$V^*_{\text{Arrhenius}} = -R \left(\frac{\partial \ln \sigma}{\partial P} \right)_T \tag{5.4}$$

Analysis based on a configurational entropy model (discussed in Chapter 6) assumes the conductivity to take the form

$$\sigma = \sigma_0 \exp - E_a/R(T-T_0) \tag{5.5}$$

where T_0 is the temperature at which the configurational entropy becomes zero and thus where translational motion effectively ceases. It is usually associated with the glass transition temperature. Assuming S_c^* and B to be independent of pressure,

$$\left(\frac{\partial K_\sigma}{\partial P} \right) = \frac{S_c^*}{RB} \left[V^* T_0 + G \left(\frac{\partial T_0}{\partial P} \right) \right] \tag{5.6}$$

and

$$\frac{V^*}{G} = \frac{T-T_0}{K_\sigma} \left[\frac{\partial \ln A}{\partial P} - \frac{\partial \ln \sigma}{\partial P} \right] - \frac{T}{T_0(T-T_0)} \left(\frac{\partial T_0}{\partial P} \right) \tag{5.7}$$

Values of V^*/G may be calculated if $\partial \ln A/\partial P$ is assumed to be zero and if $\partial T_0/\partial P$ can be estimated.

Figure 5.5 shows the effect of pressure on the conductivity for a $P(PO)_8 LiCF_3 SO_3$ sample. When fitted to a free volume equation, both E_a and T_0 are found to increase with pressure.[34] These findings lend support to a relationship between T_0 and T_g, as

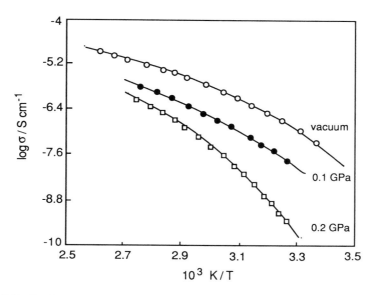

Figure 5.5. Conductivity versus reciprocal temperature for $P(PO)_8LiCF_3SO_3$ systems at different pressures. From Ref. 34.

T_g is also found to increase with pressure, and provide evidence against a liquidlike conductivity in these materials.[29]

Isothermal data such as those shown in Figure 5.6 have been obtained for various salt complexes. The data were fitted to an equation of the form $\log \sigma = \log \sigma_0 + aP = bP^2$. Zero pressure values, calculated from the slope of the conductivity versus P plot at $P = 0$, may be used to evaluate $V^*_{Arrhenius}$.

For PEO–alkali metal perchlorates and thiocyanates, analysis using Equation 5.7 and equating T_0 with T_g gave negative activation volumes. It is not clear whether this was due to an erroneous choice of T_0 or to the morphological inconsistencies in the systems studied; however, negative values were again found for V^*/G for a PPO–LiSCN complex. These results are difficult to explain and have led the authors to use the Arrhenius expression to obtain the activation volume.

Analysis in terms of Equation 5.4 provided good agreement between $V^*_{Arrhenius}$ for $P(EO)_{4.5}NaSCN$ at 40°C by Fontanella et al.[30] and by Chadwick et al.[35] $V^*_{Arrhenius}$ was found to increase with the size of both the cation and the anion in PEO–salt systems[30] but not in PPO–Li salt complexes,[34] although the values were comparable. Figure 5.7 shows a plot of activation volume versus $T - T_g$ for a number of electrolytes.[33] These include values obtained from radiotracer studies.[36] This plot suggests that the activation volume is approximately the same for all materials at a particular reduced temperature. Consequently, different activation volumes can be attributed to different T_g values. High-pressure NMR studies[33,37] also emphasize the role of T_g in ion transport. Results seemed to indicate that interchain relaxations were responsible for the ionic conductivity, as it is known that intrachain vibrations are much less affected by the pressure. It was also suggested

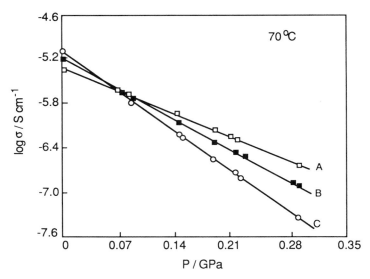

Figure 5.6. Conductivity versus pressure for (**A**) P(PO)$_8$LiSCN, (**B**) P(PO)$_8$LiCF$_3$SO$_3$, and (**C**) P(PO)$_8$LiCl4O. From Ref. 34.

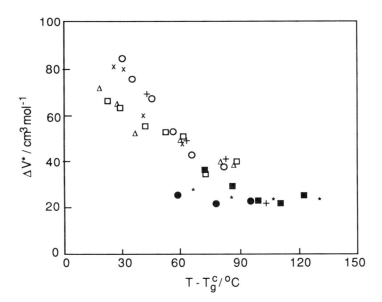

Figure 5.7. Activation volume versus temperature relative to the glass transition temperature. (■) P(PO)$_8$LiCF$_3$SO$_3$, (□) P(PO)$_8$LiClO$_4$, (△) P(PO)$_8$LiI, (×) PPO β relaxation time, (●) P(PO)$_8$LiSCN, (○) P(PO)$_8$NaClO$_4$, (+) P(PO)$_8$NaI, (*) P(PO)$_8$NaSCN. From Ref. 33. This paper was originally presented at the Fall 1987 Meeting of The Electrochemical Society, held in Honolulu, Hawaii.

that the ion sites may be in the interstices between the polymer chains, because if the ions were attached to the chains, an increase in pressure might increase the conductivity by bringing the chains closer together.

Interpreting trends in activation volumes with respect to charge transport is in fact complex. In particular, it has been proposed that factors such as the possession of a permanent dipole by the anion (e.g., trifluoromethanesulfonate or thiocyanate) may have a significant effect on its mobility[20] and thus its contribution to the conductivity.

5.3. Mixed Salt Systems

A study by Moryoussef et al.[38] showed that the conductivity of PEO–CaBr$_2$–CaI$_2$ (30:1:1) was greater than that of either PEO–CaBr$_2$ (15:1) or PEO–CaI$_2$ (15:1). A similar enhanced conductivity effect was observed when a percentage of Ca^{2+} ions were replaced by Mg^{2+} or Na$^+$ ions and was noted also by Cole[39] for PEO–ZnI$_2$–Mg(ClO$_4$)$_2$ mixtures. The effect was attributed to induced disorder in the crystalline phase; however, the conductivity enhancement was not observed when strictly anhydrous conditions were implemented. Tomlin et al.[40,41] studied lithium ion-containing mixed salt electrolytes in an attempt to improve the lithium ion conductivity. PEO–LiCF$_3$SO$_3$–NaI systems with EO:total salt content 4:1 and 8:1 were studied in detail. Mixed salt systems had associated higher conductivities, particularly at lower temperatures, and broader differential scanning calorimetry traces. Inhomogeneous distributions of anions and cations within the crystalline complex were observed, with trifluoromethanesulfonate ions aggregating near the center of the spherulites. It was concluded that the topological disorder conferred by mixing salts in PEO introduces a modulation of the stereoregularity and hence of the ion–ion and ion–polymer interactions necessary to the formation of the crystalline phase. This inherent disorder extends to give a more extensive amorphous phase and enhanced conductivity. Armand and co-workers[42] investigated mixed salt systems PEO–NaI–CuI and PEO–NaI–AgI with a view to using these as solid-state internal ionic bridges for an ionic sensor. Pure AgI is not soluble in PEO, but in the presence of excess iodide ions in the form of NaI, it dissolves most probably because iodide complexes such as AgI$_4^{3-}$ are formed. Compared with PEO–NaI electrolytes, the mixed salt systems do not show appreciable effects on the conductivity although CuI-doped systems exhibit the best values. The greatest effect is to reduce the melting temperature and to increase the activation energy above the melting point. Teeters and Norton[43] have studied PPO–KSCN–NaSCN systems to determine the possible existence of mixed alkali effects in polymer electrolytes. No clear trends were found on variation of the salt ratios, which may be a result of the rather greater contribution of the anion with respect to the cations to the conductivity.

5.4. Mixed Conductors

Mixed electronic/ionic conducting polymeric materials can be prepared by incorporating both a mobile ion and an electronic donor/acceptor species in a polymer host.

Typical examples include polyiodide materials, poly(methoxyethoxy ethoxyphosphazene)–NaI_n and $PEO-NaI_3$,[44,45] blends of polypyrrole and PEO[46,47] or polypyrrole–PEO comb copolymers,[47] and intimate $PEO-NaSCN$–polythiocyanogen $(SCN)_x$ blends.[48] Interestingly, conductivity was reported to be essentially ionic in the amorphous polyiodide system but predominately electronic in the crystalline complex. Wright[49,50] has reported PEO- or polysiloxane-based materials in which an alkali metal salt of a delocalized electron acceptor such as tetracyanoquinodimethan (TCNQ) or chloranil is coordinated. $P(EO)_3NaI$ films were exposed to a solution of TCNQ in a hydrocarbon solvent:

$$P(EO)_3NaI + TCNQ \rightleftharpoons PEO-NaTCNQ + \tfrac{1}{2} I_2$$

Following the charge transfer process, the I_2 is mostly lost to the solvent and NaTCNQ is confined to surface layers, giving a sandwich of PEO–NaI between electronically conducting surfaces. Conductivity is found to be largely electronic, with electron transfer taking place along the TCNQ stack normal to the plane of the radical anions. The conductivity of the surface layers is critically dependent on the polymer electrolyte morphology which is consistent with the formation of a molecular complex. Exposure of films to I_2 vapor[49] brings about the formation of a $NaTCNQ-I_2$ complex salt with a hundredfold increase in conductivity to approximately 1 S cm^{-1} at room temperature. The presence of iodine atoms in the lattice is assumed to give rise to back-oxidation of TCNQ radical anions, creating holes within the stack and facilitating electron transfer along it.

The materials are interesting because of their mixed conduction mechanism and their possible applications in electrochemical devices, such as assisting interfacial ion transport between electrolyte and electrodes, but detailed studies of, for example, the importance of electronic versus ionic conductivity, morphology, or ion–ion and ion–polymer interactions, have not been fully elucidated.

References

1. J. R. MacCallum, M. J. Smith, and C. A. Vincent, *Solid State Ionics* **11** (1984), 307.

2. F. M. Gray, *Solid State Ionics* **40/41** (1990), 637.

3. M. Stainer, L. C. Hardy, D. H. Whitmore, and D. F. Shriver, *J. Electrochem. Soc.* **131** (1984), 784.

4. M. B. Armand, J. M. Chabagno, and M. Duclot, in *Fast Ion Transport in Solids* (P. Vashista, J. N. Mundy, and G. K. Shenoy, Eds.), Elsevier/North-Holland, New York (1979), p. 131.

5. R. Dupon, B. L. Papke, M. A. Ratner, and D. F. Shriver, *J. Electrochem. Soc.* **131** (1984), 586.

6. D. W. Xia, D. Soltz, and J. Smid, *Solid State Ionics* **14** (1984), 221.

7. P. M. Blonsky, D. F. Shriver, P. Austin, and H. R. Allcock, *Solid State Ionics* **18/19** (1986), 258.

8. A. Killis, J. F. Le Nest, A. Gandini, and H. Cheradame, *J. Polym. Sci. Polym. Phys. Ed.* **19** (1981), 1073.

9. C. K. Chiang, G. T. Davis, C. A. Harding, and T. Takahashi, *Macromolecules* **18** (1985), 825.

10. S. Fang, P. Zhang, and Y. Jiang, *Polym. Bull.* **19** (1988), 81.

11. F. M. Gray, J. R. MacCallum, C. A. Vincent, and J. R. M. Giles, *Macromolecules* **21** (1988), 392.

12. J. E. Weston and B. C. H. Steele, *Solid State Ionics* **2** (1981), 347.

13. J. E. Weston and B. C. H. Steele, *Solid State Ionics* **7** (1982), 81.

14. F. M. Gray, C. A. Vincent, and M. Kent, *J. Polym. Sci. Polym. Phys. Ed.* **27** (1989), 2011.

15. W. Gorecki, R. Andreani, C. Berthier, M. B. Armand, M. Mali, J. Roos, and D. Brinkmann, *Solid State Ionics* **18/19** (1986), 295.

16. C. Berthier, W. Gorecki, M. Minier, M. B. Armand, J. M. Chabagno, and P. Rigaud, *Solid State Ionics* **11** (1983), 91.

17. M. Minier, C. Berthier, and W. Gorecki, *J. Phys.* **45** (1984), 739.

18. R. Landauer, *J. Appl. Phys.* **23** (1952), 779.

19. S. Kirkpatrick, *Rev. Mod. Phys.* **45** (1973), 574.

20. Y. L. Lee and B. Crist, *J. Appl. Phys.* **60** (1986), 2683.

21. A. R. P. Le Mehaute and G. Crepy, *Solid State Ionics* **9/10** (1983), 17.

22. A. R. P. Le Mehaute, in *Proceedings, 6th Risø International Symposium on Metallurgy and Materials Science* (F. W. Poulsen, N. Hassel Andersen, K. Clausen, S. Skaarup, and O. T. Sørensen, Eds.), Risø National Lab., Roskilde (1985).

23. T. Hamaide, A. Guyot, A. Le Mehaute, G. Crepy, and G. Marcellin, *J. Electrochem. Soc.* **136** (1989), 3152.

24. J. F. Le Nest, A. Gandini, and H. Cheradame, *Br. Polym. J.* **20** (1988), 253.

25. J. R. MacCallum, A. S. Tomlin, and C. A. Vincent, *Eur. Polym. J.* **22** (1986), 787.

26. W. L. Archer and R. D. Armstrong, *Electrochim. Acta* **25** (1980), 1689.

27. C. A. Vincent, *Prog. Solid State Chem.* **17** (1987), 145.

28. C. P. Flynn, *Point Defects and Diffusion*, Clarendon Press, Oxford (1972).

29. J. J. Fontanella, M. C. Wintersgill, M. K. Smith, J. Semancik, and C. G. Andeen, *J. Appl. Phys.* **60** (1986), 2665.

30. J. J. Fontanella, M. C. Wintersgill, J. P. Calame, F. P. Pursel, D. R. Figueroa, and C. G. Andeen, *Solid State Ionics* **9/10** (1983), 1139.

31. S. G. Greenbaum, Y. S. Pak, M. C. Wintersgill, J. J. Fontanella, J. W. Schulz, and C. G. Andeen, *J. Electrochem. Soc.* **135** (1988), 235.

32. M. C. Wintersgill, J. J. Fontanella, M. K. Smith, S. G. Greenbaum, K. J. Adamic, and C. G. Andeen, *Polymer* **28** (1987), 633.

33. S. G. Greenbaum, K. J. Adamic, Y. S. Pak, M. C. Wintersgill, J. J. Fontanella, D. A. Beam, and C. G. Andeen, in *Proceedings, Electrochemical Society Symposium on Electro-ceramics and Solid State Ionics, Honolulu, Hawaii, 1987* (H. Tuller, Ed.), Electrochem. Soc., Pennington, N.J. (1988).

34. J. J. Fontanella, M. C. Wintersgill, J. P. Calame, M. K. Smith, and C. G. Andeen, *Solid State Ionics* **18/19** (1986), 253.

35. A. V. Chadwick, J. H. Strange, and M. R. Worboys, *Solid State Ionics* **9/10** (1983), 1155.

36. C. Bridges and A. V. Chadwick, *Solid State Ionics* **28–38** (1988), 965.

37. M. C. Wintersgill, J. J. Fontanella, S. G. Greenbaum, and K. J. Adamic, *Br. Polym. J.* **20** (1988), 195.

38. A. Moryoussef, M. Bonat, M. Fouletier, and P. Hicter, in *Proceedings, 6th Risø International Symposium on Metallurgy and Materials Science* (F. W. Poulsen, N. Hassel Andersen, K. Clausen, S. Skaarup, and O. T. Sørensen, Eds.), Risø National Lab., Roskilde (1985), p. 335.

39. M. Cole, Ph.D. thesis, Leicester Polytechnic (1989).

40. J. R. MacCallum, A. S. Tomlin, D. P. Tunstall, and C. A. Vincent, *Br. Polym. J.* **20** (1988), 203.

41. D. P. Tunstall, A. S. Tomlin, F. M. Gray, J. R. MacCallum, and C. A. Vincent, *J. Phys, Condens. Mater.* **1** (1989), 4035.

42. P. Fabry, C. Montero-Ocampo, and M. Armand, *Sensors Actuators* **15** (1988), 1.

43. D. Teeters and J. Norton, *Solid State Ionics* **40/41** (1990), 648.

44. M. M. Lerner, L. J. Lyons, J. J. Tonge, and D. F. Shriver, *Polym. Prepr.* **30** No. 1 (1989), 435.

45. J. A. Siddiqui and P. V. Wright, *Faraday Discuss. Chem. Soc.* **88** (1989), 113.

46. O. Inganäs, *Br. Polym. J.* **20** (1988), 233.

47. M. G. Minett and J. R. Owen, *Solid State Ionics* **28–30** (1988), 1192.

48. K. Conheeney and P. V. Wright, *Polym. Commun.* **27** (1986), 364.

49. B. Mussarat, K. Conheeney, J. A. Siddiqui, and P. V. Wright, *Br. Polym. J.* **20** (1988), 293.

50. J. A. Siddiqui and P. V. Wright, *Polym. Commun.* **28** (1987), 90.

CHAPTER 6

Polymer Electrolyte Architecture

Polymer electrolytes that are based on high-molecular-weight poly(ethylene oxide) (PEO) are themselves of limited applicability in the development of practical devices because of a number of restrictive properties inherent in the polymer. High-molecular-weight PEO is highly crystalline and salt complexes tend to contain either crystalline PEO and/or a high-melting polymer–salt complex. Nuclear magnetic resonance (NMR) studies carried out by Berthier et al. have shown conclusively that ionic motion in polymer electrolytes is predominately through the amorphous phase.[1] At ambient temperatures, PEO-based electrolytes are generally poor conductors ($< \sigma \sim 10^{-8}$ S cm^{-1}) due to the high degree of crystallinity. A number of salts such as RbSCN, RbI, CsSCN, CsI, and Hg(ClO$_4$)$_2$,[2-4] do form low-temperature amorphous polymer electrolyte systems with PEO, but with major interest centered around producing lithium-based electrochemical devices, an amorphous polymer–lithium salt electrolyte is desirable. Reasonable conductivity in PEO–salt polymer electrolytes can be achieved ($\sim 10^{-5}$ S cm^{-1}) at temperatures of the order of 100°C, but the mechanical properties of the polymer are significantly poorer. The loss of mechanical stability is largely the result of the melting of the crystalline phases and it is the absence of crystallinity in the polymer hosts, described later, that often gives them their inferior mechanical strength, when compared with PEO systems.

6.1. Non-Ether-Based Polymer Electrolytes

A number of polymer materials have been investigated as alternatives to polyethers as salt coordinating systems. The majority have been based on polyesters although a number based on a flexible siloxane structure are discussed later.

Studies of polymer electrolytes formed between poly(ethylene adipate) (PEA), [—OCH$_2$CH$_2$OOC(CH$_2$)$_4$CO—]$_n$, of different molecular weights and LiCF$_3$SO$_3$[5] and NaI[6] have been reported. The polymer is similar to PEO in that the high-molecular-weight material has a T_g of −50°C and a T_m of 55°C. Conductivities of

$LiCF_3SO_3$ systems were reported to be better than those for analogous PEO electrolytes although mechanical properties were poor above the melting point. NaI systems formed poor ionic conductors. A number of studies have been reported for poly(ethylene succinate), $[—OCH_2CH_2OCOCH_2CH_2O—]_n$,[7–10] incorporating $LiBF_4$ and $LiClO_4$. Poor conductivities were found for these materials as was the case with another polyester, poly(ethylene sebacate), $[—OCH_2CH_2OCO(CH_2)_8-CO—]_n$, where unfavorable coordinating group spacing makes intramolecular interactions difficult.[8] Poly(β-propiolactone), $[—CH_2CH_2\ COO—]_n$, coordinated with $LiClO_4$[11] is semicrystalline, with the degree of crystallinity decreasing with increasing salt content. Conductivities were again poor but showed an increase with increasing salt content over the range 100:1 to 10:1 polymer repeat unit-to-lithium. Metastable quenched samples exhibited a conductivity of 3.7×10^{-4} S cm^{-1} at 70°C. Unlike polyether and related systems, many of the alternative polymers are unlikely to see further development partly because they have no advantages with respect to conductivity, but also because their poorer chemical stability makes them unfavorable for battery development.

6.2. Amorphous Polyether-Based Polymer Architecture

Using an all-amorphous polymer is one way of enhancing the ionic conductivity of a polymer electrolyte. It is necessary, when redesigning the polymer architecture, to incorporate the optimal distribution of solvating groups for cation coordination and to minimize chain regularity, from the point of view of reducing crystallinity. Polyethylene glycols of molecular weight less than 500 are amorphous liquids and

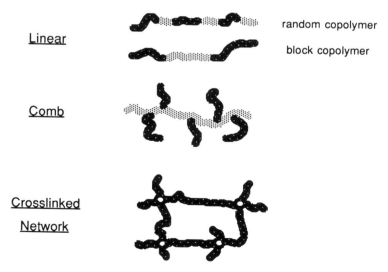

Figure 6.1. Various polymer host architectures for reduced crystallinity.

are therefore of little value as polymer electrolyte hosts themselves, but, by incorporating these oligomers into a high-molecular-weight polymer structure where they act as the complexing sites for ions, numerous amorphous architectures of varying properties may be designed. In recent years many new polymers of this form (and some completely new classes of polymer) have been reported that are aimed at optimizing the electrical and mechanical properties of polymer electrolytes. The amorphous polymers are normally based on one or more of the structures shown in Figure 6.1.

6.3. Enhanced Chain Flexibility

Polymer electrolyte materials of minimized T_g and optimum ionic conductivity may be produced by introducing a highly flexible copolymeric unit into the structure. Polyphosphazene ($T_g = -70°C$) and polydimethylsiloxane ($T_g = -123°C$) have been used as the backbone for a number of copolymers that are found to solvate ions and exhibit ionic conductivities two to three orders of magnitude higher than those of the analogous PEO-based material. In particular, a polyphosphazene-based system[12] has been successfully used in a number of research laboratories for various electrical studies.[13–17]

Shriver and co-workers[12,18,19] synthesized a range of polyphosphazene-based polymers of the general formula

$$
\begin{array}{c}
O-(CH_2CH_2O)_nCH_3 \\
| \\
-(N=P-)_m \\
| \\
O-(CH_2CH_2O)_nCH_3
\end{array}
$$

with variable n values. These were formed by substituting the chlorine atoms of poly(dichlorophosphazene) using the sodium salt of 2-(2-methoxyethoxy)ethanol in the presence of tetra-n-butylammonium bromide. The most successful material has been the polymer with $n = 2$, poly(methoxyethoxy ethoxyphosphazene) (MEEP).[12] Perhaps not surprisingly, the dimensional stability of MEEP is poor and therefore the properties are undesirable for many applications. Two effective crosslinking methods for this material have been reported that lead to minimal loss in ionic conductivity. Tonge and Shriver[20] chemically crosslinked MEEP by preparing the polymer as described earlier but substituting a proportion of the methoxyethoxy ethanol with poly(ethylene glycol) (PEG), $HO(CH_2CH_2O)_nOH$. PEG at 1 mol% was sufficient to markedly increase the dimensional stability: linear MEEP will flow under a constant pressure, particularly above 70°C, whereas the crosslinked material was stable, even at 140°C. The conductivities were reported to be comparable to those of the analogous uncrosslinked systems. Recently,[21,22] crosslinked MEEP has been prepared by ^{60}Co gamma irradiation. Radiation dose rates of up to 20 Mrad h^{-1} were used. The crosslinking was found to occur through the ethylene oxide units, leaving the flexible backbone unaffected, which is similar to the chemical crosslinked material described earlier. Conductivities were unchanged by the degree of crosslinking and did not vary by more than 5×10^{-5} S cm^{-1} from those of an uncrosslinked sample. A similar radiation-crosslinked polyphosphazene has been reported by Nazri and Meibuhr.[23]

A number of studies of siloxane-based polymer electrolytes have been reported in the literature. Nagaoka et al.[24] prepared a linear polymer, poly(dimethyl siloxane-co-ethylene oxide) (DMS–nEO), by reacting dimethyldichlorosilane with short-chain polyethylene glycols:

$$\begin{array}{ccccc} & CH_3 & & & CH_3 \\ & | & & & | \\ Cl—Si—Cl & + & HO—(CH_2CH_2O)_nH & \rightarrow & —(Si O—(CH_2CH_2O)_n—)_m + HC \\ & | & & & | \\ & CH_3 & & & CH_3 \end{array}$$

$$\text{DMS-}n \text{ EO}$$
$$(n = 1,2,4,9)$$

The highest conductivity at 25°C, 1.5×10^{-4} S cm^{-1}, was found for a LiClO$_4$–DMS–4EO electrolyte.

Hall et al.[25] and Smid and co-workers[26–28] produced liquid comb-branched poly{[ω-methoxyoligo(oxyethylene)ethoxy]methyl siloxane}s (PMMS-m) by the reaction

$$\begin{array}{ccccc} CH_3 & & & & CH_3 \\ | & & & \text{catalyst} & | \\ —(Si O)_n— & + & HO—(CH_2CH_2O)_nCH_3 & \longrightarrow & (Si O)_n + H_2 \\ | & & & & | \\ H & & & & O—(CH_2CH_2O)_nCH_3 \end{array}$$

$$\text{PMMS-}m$$

with $n = 5,7,8,12,16,$ and 22 and $m = n + 1$. Zinc octanoate[26] and triethylamine[25] were used as catalysts. The length of the side chain did not appear to greatly affect the conductivity, which was of the order of 1×10^{-4} S cm^{-1} for LiClO$_4$ electrolytes at room temperature, although side-chain crystallization was observed for $n = 12–22$. The polymers could be crosslinked by heating to 150°C or at a lower temperature with benzoyl peroxide initiator. Crosslinking reduced the conductivities by an order of magnitude. Alternatively, the crosslink density could be chemically controlled by substituting a percentage of the monomethoxy poly(ethylene glycol) with poly(ethylene glycol) in the reaction sequence.[27]

The susceptibility of Si—O—C bonds to hydrolysis and subsequent structural degradation is a severe problem unless moisture can be completely eliminated. Si—C bonds, on the other hand, are very stable and copolymers containing this type of linkage are therefore more desirable. The synthesis and properties of comb-branched polysiloxanes, poly{[ω-methoxyoligo(oxyethylene)propyl]methyl siloxane}s (PAGS-m),[28] having the structure

CH$_3$
|
$\dashv$Si O$\dashv_x$
|
(CH$_2$)$_3$
|
O$\dashv$CH$_2$CH$_2$O)$_n$CH$_3$

PAGS-*m*

have been reported. These were prepared by reacting poly(methyl siloxane), [$-$(Me)HSiO$-$]$_n$, and the allyl ether CH$_2$=CHCH$_2$O(CH$_2$CH$_2$O)$_n$CH$_3$, n = 2,7,9, and 12, in the presence of a platinum complex catalyst.[28–30] Eighty percent of the Si-H groups were found to react. Ionic conductivities of LiClO$_4$ complexes were very similar to those of the PMMS series.[27] A similar preparation was carried out to produce a polyelectrolyte derivative of this polymer[31,32]:

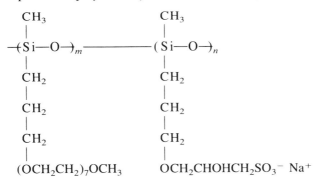

A percentage of the allyl ether was replaced by an allyl glycidyl ether, CH$_2$=CHCH$_2$OCH$_2$CH(O)CH$_2$. The epoxy ring underwent further reaction with Na$_2$S$_2$O$_5$ to give the sulfonate end group. The anion is thus incorporated as part of the polymer and immobilized. Increasing the sulfonate content from 0 to 70% produced materials with properties ranging from liquids to waxes to solids. Skotheim and co-workers[33] have reported on siloxane-based polyelectrolytes where there is steric hindrance of the anionic group attached to the polymer backbone, to prevent charge association. With sterically hindered phenol compounds, conduc-.tivities of the order of 10^{-5} to 10^{-7} S cm^{-1} were achieved.

The T_g of polyurethane-linked polyether networks may also be reduced to enhance their conductivity by incorporating poly(dimethylsiloxane) units into the polymer framework.[34–36] The basic reaction for preparing PEO–polyurethane networks is

HO$\dashv$CH$_2$CH$_2$O$\rightarrow_n$H + R$-$N=C=O $\rightarrow$ R$-$N$-$C=O
poly(ethylene glycol)   isocyanate | |
                               (crosslinking agent)   H   O$\dashv$CH$_2$CH$_2$O$-$)$_n$H

                                                                        polyurethane unit

PEO can be replaced by the grafted copolymer

$$
\begin{array}{ccc}
 & CH_3 & CH_3 \\
 & | & | \\
(CH_3)_3-SiO\!+\!SiO\,\}_x\!+\!SiO\,\}_y\!-\!Si-(CH_3)_3 \\
 & | & | \\
 & CH_3 & (CH_2)_3 \\
 & & | \\
 & & (CH_2CH_2O)_zH
\end{array}
$$

to produce the desired network. Copolymer molecular weights were typically about 20,000 and $z = 20$ or 40. Synthetic details are considered by Lestel et al.[35] Conductivities of the order of 10^{-5} to 10^{-6} S cm^{-1} have been reported for LiClO$_4$-network electrolytes and thermal stability studies showed that no creep was observed up to 120°C. The crosslinked siloxane structure

$$
\begin{array}{ccc}
 & CH_3 & CH_3 \\
 & | & | \\
PEO\!-\!(CH_2)_3\!-\!Si\!-\!O\!-\!Si\!-\!(CH_2)_3\,PEO \\
 & | & | \\
 & O & O \\
 & | & | \\
PEO\!-\!(CH_2)_3\!-\!Si\!-\!O\!-\!Si\!-\!(CH_2)_3\,PEO \\
 & | & | \\
 & CH_3 & CH_3
\end{array}
$$

has also been investigated by this group.[37,38] Watanabe et al.[39] prepared a polyurethane network using a different concentration and length of ether chains in the preceding copolymer. Crosslinked networks based on siloxane–ethylene oxide block copolymers have been prepared.[40] An ethylene oxide–dimethyl siloxane AB or ABA block copolymer was chain extended and complexed with the appropriate salt before it was crosslinked by a condensation reaction with 3-isocyanatopropyltriethoxysilane.

New classes of ionically conducting polymer based on a crosslinked siloxane structure have been reported by Fang et al.[41] and Wnek et al.[42] The former consists of a methylsiloxane backbone with a polymeric trifunctional siloxane as the crosslinking agent. Polymers have the form

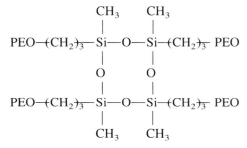

The flexibility of the siloxane backbone allows variable orientations of the polar cyano groups, which thus successfully solvate salts. By optimizing parameters such as crosslinking group and polar group density, network films with conductivities close to 10^{-5} S cm^{-1} may be realized at room temperature. Similarly, a conducting polymer of the form

$$
\begin{array}{c}
CH_3 \\
| \\
-(Si-O)_n- \\
| \\
(CH_2)_3 \\
| \\
CH_3-C-S-\text{\Large\textcircled{}}-NO_2 \\
| \\
H_3CO-C \\
\| \\
O
\end{array}
$$

has recently been reported.[42] These are viscous liquids that may be crosslinked and are able to dissolve large quantities of salts such as tetra-n-butylammonium chloride.

A further crosslinked siloxane-based polymer electrolyte worth noting is that produced by the plasma polymerization of tris(2-methoxyethoxy)vinyl silane.[43–46] After polymerization conditions were optimized, conductivities of the order of 5 × 10^{-5} S cm^{-1} were reported. Sulfonate groups may also be introduced prior to polymerization.[44,45] The lithium sulfonate polymer (0.1–0.3% end groups) blended with 10% PEO(300) exhibited an ionic conductivity of 10^{-6} S cm^{-1} at 60°C.

Cowie et al.[47] prepared polymers with a sterically unhindered backbone to maximize chain flexibility. The vinyl monomer CH_2=$CHO(CH_2CH_2O)_3CH_3$ (VMEO$_3$) was polymerized using borontrifluoride etherate as initiator. The isolated polymer had a molecular weight of 30,000 and a T_g of − 66.5°C. Ambient temperature conductivities were of the order of 10^{-4} S cm^{-1} for a 20:1 polymer–LiClO$_4$ complex. A crosslinked analog has also been reported[48,49]:

$$
\begin{array}{c}
-(CH-CH_2)_{n-x}-(CA)_x- \\
| \\
O \\
| \\
(CH_2CH_2O)_2CH_2CH_3
\end{array}
$$

where CA represents the crosslinking group. The ether side groups were either ethoxy or methoxy terminated in this instance. Room temperature conductivities of 10^{-6} S cm^{-1} for an 8:1 LiClO$_4$ complex and 10^{-5} S cm^{-1} for a LiBF$_4$ complex were found but the mechanical properties were much improved.

Ballard et al.[30] prepared an extensive range of comb-branched copolymers based on the general structure

$$-\negthickspace\left[CH_2CH_2O\right]_x\negthickspace\left[CH_2CHO\right]_y\negthickspace-$$
$$|$$
$$R$$

where the side group R was an ether of varying chain length and end group. In addition to the already established salt concentration and T_g effects, room temperature ionic conductivity was found to be optimized for polymers with a carbon to oxygen ratio close to 2:1.

6.4. Mechanical Stability

6.4.1. Networks

The previous section highlighted the dimensional stability problem of decreasing the T_g of host polymers to enhance ionic conductivity. This can be overcome largely by crosslinking the material. Both radiation and chemical crosslinking have been used extensively to produce amorphous, mechanically stable networks. In particular, Cheradame and co-workers[34,36,50-64] have concentrated on forming chemically crosslinked polyurethane structures. The basic reaction involves the addition of an appropriate salt and crosslinking agent, usually an isocyanate, to a multifunctional oligomer such as poly(ethylene glycol), poly(propylene glycol), glycol ABA triblock copolymers, polyether triol, or more complex structures such as the siloxane–ether copolymers described earlier. The resulting mixtures are cured at elevated temperatures. Provided the molecular weight between crosslinks is less than 1500, PEO-based networks show no tendency to crystallize. It has been suggested[62] that the optimum conductivity of PEO networks corresponds to a system with PEO units of molecular weight $1000n$, containing $2n$-1 mol of lithium ions per mole of PEO unit. The crosslink density affects the conductivity by restricting segmental motion. A linear relationship of $1/T_g$ with the crosslink density and with salt concentration has been observed where

$$T_g^{-1} = T_{g0}^{-1} - 7.6 \times 10^{-4}\, c$$

for the salt-free polymer and

$$T_g^{-1} = T_{g0'}^{-1} - 2.7 \times 10^{-4}\, c'$$

for the complexed networks. c is the crosslink concentration, T_{g0} is the glass transition temperature at $c = 0$, c' is the salt concentration, and $T_{g0'}$ is the glass transition temperature of the salt-free network. These observations allowed the construction of a diagram (Figure 6.2) from which the T_g of any LiClO$_4$-containing PEO-based network, at any concentration, could be determined. The crosslinking agent was found not to affect the slope. This may be explained by the fact that the free volume of the network is a function only of the interactions between the salt and the polymer chains. However, replacing PEO with a PEO–poly(propylene oxide) (PPO) block gave an increased slope on the graph. As complexation of Li$^+$ ions is much more favored by PEO, a recalculation of the concentration gave a slope close

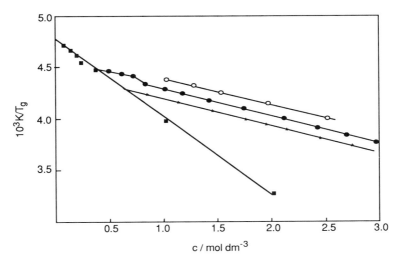

Figure 6.2. Plot of reciprocal T_g for PEO networks versus overall concentration of urethane networks (■) and LiClO$_4$ for PEO(1000) (*), PEO(2000) (•), and PEO(3000) (○). From Refs. 62 and 64.

to 2.7×10^{-4} dm^3 mol^{-1} K^{-1}.[63] The effect would be expected to be even more complex in the case of a siloxane-based copolymer, as the siloxane unit does not solvate cations at all, yet reduces the T_g of the final network. For PPO-based networks, the urethane linkage has a greater influence.[36] The oxygen atom in PPO is more nucleophilic than the corresponding oxygen in PEO and may possibly hydrogen bond more strongly with the linking group. Dynamic mechanical properties of network electrolytes have been investigated[53,58] and it is found that different modulus plots could be superimposed on a reduced temperature scale, independent of network structure and salt content. Watanabe and co-workers have also prepared polyurethane networks, using PPO diols and triols with an isocyanate or sebacoyl chloride as linking agent.[65–71] Electrolyte was added, in this instance, by swelling the polymer in a salt solution.

The incorporation of isocyanate units into the network polymer results in a polymer electrolyte containing large quantities of bulky groups, which are superfluous to the conduction mechanism and may in fact hinder ionic motion. In addition, the urethane linkage has a strong influence on the T_g because of hydrogen bonding and van der Waals interactions between them and the surrounding polyether. An improved amorphous network system based on more flexible phosphate ester crosslinks of PEGs may be brought about by the reaction[72,73]

$$3H(OCH_2CH_2)_nOH + 2POCl_3 \rightarrow [(OCH_2CH_2)_nO]_3 [PO]_2 + 6HCl$$

Conductivities of approximately 2×10^{-5} S cm^{-1} were achieved when LiCF$_3$SO$_3$

was incorporated into the polymer. Ballard et al.[30] have described the preparation of a crosslinked network based on the structure

$$-(CH_2CHO)_{\overline{x}}(CH_2CH_2O)_{\overline{y}}(CH_2CHO)_{\overline{z}}-$$
$$\quad\ \ |\qquad\qquad\qquad\qquad\qquad\ \ |$$
$$\quad\ \ CH_2\qquad\qquad\qquad\qquad\quad CH_2$$
$$\quad\ \ |\qquad\qquad\qquad\qquad\qquad\ \ |$$
$$\quad\ \ O(CH_2CH_2O)_2CH_3\qquad\quad OCH_2CH{=}CH_2$$

Films were cast from a solution of copolymer, $LiCF_3SO_3$, and benzoyl peroxide and the unsaturated centers were subsequently crosslinked by heating under vacuum at $110°C$. It was reported that the crosslinked material had a lower T_g than the uncrosslinked copolymer but a conductivity an order of magnitude lower.

High-molecular-weight PEO complexes are amorphous at elevated temperatures, with the actual temperature range dependent on the salt and its concentration. After being heated, lithium perchlorate complexes supercool and retain the amorphous state for several days. The recrystallization of samples of $P(EO)_8LiClO_4$ could be inhibited indefinitely by lightly crosslinking such supercooled materials using gamma radiation from a ^{60}Co source.[74] The conductivity was found to be only slightly impaired. By contrast, crosslinking films of $P(EO)_9LiCF_3SO_3$ at $78°C$ could not prevent recrystallization, although evidence suggested that the uncomplexed PEO in the electrolyte was in fact crosslinked.[75] It would appear that in the latter case, the polymer chains have not undergone appreciable randomization, thus allowing the chains to be crosslinked in a position that favors crystallization. In addition, $P(EO)_9LiCF_3SO_3$ is only partially amorphous at $78°C$.

Le Mehaute et al.[76,77] improved the long-term performance of cells by producing a rather complex network system that retained high ionic conductivity over many years. PEO, styrene-terminated PEO, and a butadiene–acrylonitrile copolymer were mixed together and subjected to X-ray or thermal treatment that polymerized and crosslinked the multicomponent system. Polyglycerine, an ether oxygen-containing polymer, was used as the backbone for a pendant ether comb-type polymer.[78] The polymer was complexed with $LiClO_4$ and crosslinked to give mechanical stability. The chain length and ratio of EO to PO determined the conductivity, with room temperature values of $\sim10^{-5}$ S cm^{-1} for amorphous systems reported.

6.4.2. Comb-Branched Copolymers

Comb-branched systems in which the backbone of a low-T_g polymer imparts enhanced flexibility to the system have already been described. Alternatively, the backbone may be of a high T_g to give the system greater mechanical stability. A number of systems based on substituted methacrylate polymers have been described in the literature[79-83]:

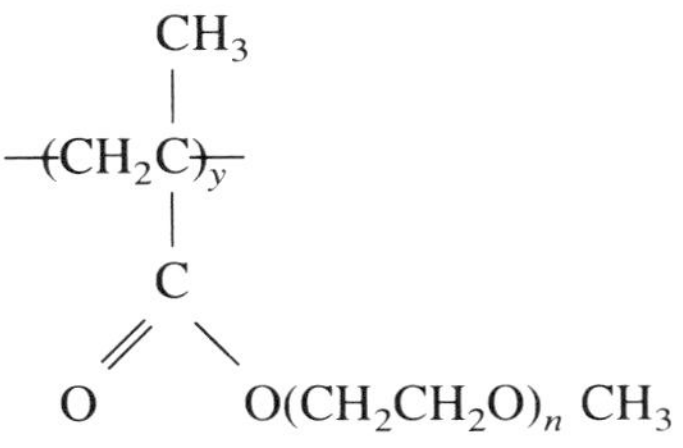

$$CH_3$$
$$|$$
$$-(CH_2C)_y-$$
$$|$$
$$C$$
$$/\!\!/ \quad \backslash$$
$$O \qquad O(CH_2CH_2O)_n\ CH_3$$

PMG(n)

Here $n = 7-22$. In addition to the value of n, the type and concentration of salt affected the formation of a crystalline complex.[80] A similar system but with a much higher PEO oligomer side-chain density is that based on a polyitaconate backbone[84-86]:

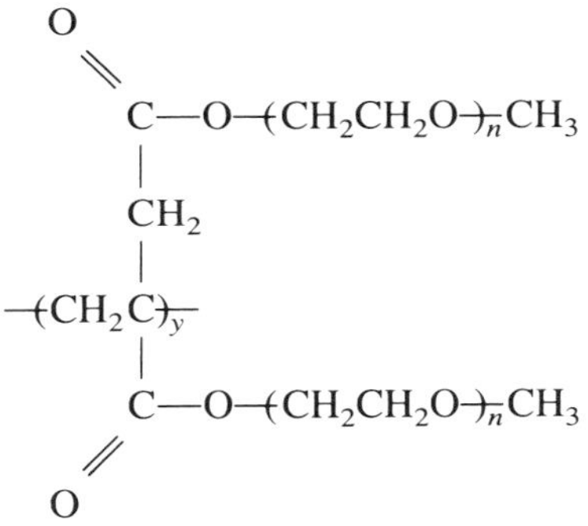

$$O$$
$$\backslash\!\backslash$$
$$\quad C-O-(CH_2CH_2O)_n CH_3$$
$$\quad |$$
$$\quad CH_2$$
$$\quad |$$
$$-(CH_2C)_y-$$
$$\quad |$$
$$\quad C-O-(CH_2CH_2O)_n CH_3$$
$$\quad /\!\!/$$
$$O$$

PEO(n)MI

Here $n = 1-5$. These polymers and their complexes were found to be wholly amorphous up to O:M$^+$ ratio of 2:1. A similar series with PPO oligomer side chains was prepared by Cowie and co-workers.[86-88] Samples with $n = 6$ and $n = 17$ were amorphous polymers forming electrolytes with LiClO$_4$ or NaClO$_4$, which had conductivities comparable to or higher than those of the PEO analogs. A rigid poly(γ-methyl-L-glutamate) main chain polymer with ethylene oxide pendant groups that dissolves various salts has been prepared[89]:

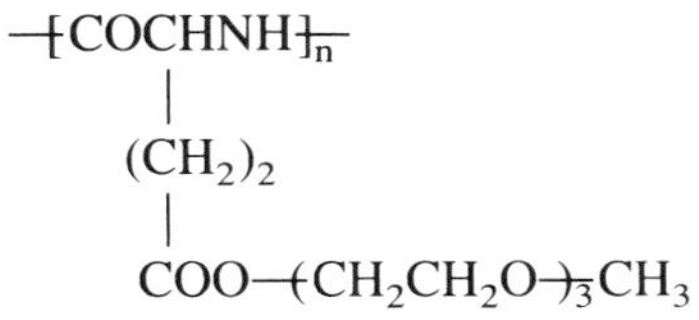

$$-(COCHNH)_n-$$
$$|$$
$$(CH_2)_2$$
$$|$$
$$COO-(CH_2CH_2O)_3CH_3$$

The conductivity was reported to increase with decreasing lattice energy of the dissolved salt.

6.4.3. Block Copolymers

Amorphous polyurethane networks formed from PEO–PPO–PEO triblock copolymers have been studied by Cheradame.[59,62] The advantage of the triblock over a linear polyether in the network synthesis is the reduction of the polyurethane linkage density, which adversely affects the T_g of the material.

Watanabe and co-workers[90] studied an $(-AB-)_n$-type system, polyether poly(urethane urea) (PEUU):

PEUU constitutes 30% by weight of the total and is the *hard* segment; the polyether is the *soft* segment. The hard segments tend to aggregate as a result of hydrogen bonding so that a microphase separated structure is established that gives the material excellent structural properties. Salts dissolved selectively in the polyether phase, which provided a continuous conduction path for the ions when its volume fraction was above the percolation limit. Correlation between the morphology, ionic conductivity, and dynamic mechanical properties has been described.[70,71,90] Watanabe et al.[91] used the physical properties of polyions to solidify low-molecular-weight PPO:

A poly-(tetramethylene oxide) polymer was also synthesized. The regular charge distribution down the chain imparts rigidity to the structure. Conductivities were an order of magnitude lower than for a PPO homopolymer. This may be due in part to the higher T_g −43.2°C for PPO units of molecular weight 3000, compared to −60°C for the homopolymer) but interactions of the salt with charges on the backbone are also likely.

A sodium thiocyanate complex with a poly(b-ethylene oxide–b-isoprene–b-ethylene oxide) ABA block copolymer was studied by Robitaille and Prud'homme.[92] The PEO end blocks tended to crystallize and were shown to form finely dispersed isolated microdomains. The morphological properties of ABA block copolymers instilled good dimensional stability in the material, even at high temperature.

A novel type of polymer architecture that uses the properties of block copolymers to optimize ionic conductivity while retaining excellent mechanical properties has been described by Gray et al.[93,94] Short-chain PEO segments were grafted onto the flexible central segment of a styrene–butadiene–styrene ABA block copolymer. The hard "A" phase is incompatible with the elastomeric "B" portion, and by dissolving the polymer and salt in a suitable mixed solvent, microphase separation occurs to provide reinforcement sites within the continuous, highly mobile conducting matrix. The materials were best described as self-supporting rubbers and conductivities of approximately 10^{-5} S cm^{-1} could be achieved at room temperature.

6.4.4. Random Polyethers

Crystallinity can be reduced by incorporating secondary units into the PEO chain. If these are regularly spaced, long EO sequences do not exist and the polymer melting point is accordingly depressed. Foos and Erker[95] and Armand and co-workers[89] prepared polydioxolane, $—(CH_2CH_2OCH_2O)_n—$, by chemical and electrochemical polymerization. As LiAsF$_6$ catalyzes the polymerization of dioxolane, polymer electrolytes could be prepared in situ. Chemical oxidation of different concentrations of LiAsF$_6$ in dioxolane was initiated by dichlorodicyanobenzoquinone. The mixture polymerized at room temperature. Electrochemical polymerization was initiated by applying a current or voltage pulse between stainless-steel electrodes. At 3.0 V versus Li or 0.3 mA cm^{-2}, the dioxolane was oxidized at the anode and Li plated at the cathode. After initiation, the cell was allowed to stand at open circuit for varying lengths of time, depending on the thickness of film required. The polymer was found to have a broad molecular weight distribution with $m_w \sim 5 \times 10^4$.[96] The crystalline material melted at 55°C and had a T_g of -65°C. Complexes with LiCF$_3$SO$_3$ and Li(CF$_3$SO$_2$)$_2$N were amorphous at room temperature with conductivities of 10^{-5} to 10^{-6} S cm^{-1} at O:Li ratios of 8:1 to 16:1. Electrochemical and chemical preparation methods gave materials with comparable conductivities.[95] Higher polymerization temperatures favored macrocyclic ring formation, which may be an impurity in these systems. In addition, because of the poorer solvating properties of polydioxolane in comparison with PEO, there may be a greater tendency for cations to hold solvent molecules which subsequently affects the properties of the polymer electrolyte.

A similar mixed oxa-alkane, described by Giles et al.,[97] was formed by copolymerization of trioxane and dioxolane in such a way as to give a random distribution of oxyethylene and oxymethylene units:

Polymerization was initiated using boron trifluoride etherate, $BF_3(C_6H_9)_2O$. The product was a viscous melt with molecular weight 23,000. Differential scanning calorimetry (DSC) and conductivity data for $LiCF_3SO_3$ complexes with the polymer indicate the presence of a crystalline copolymer that melts around room temperature and a high-melting crystalline complex. Room temperature conductivities were of the order of 1 to 5×10^{-6} cm^{-1} for O:Li ratios in the range 9:1 to 35:1.

The synthesis of a more successful oxyethylene–oxymethylene structure in which EO units longer than those described earlier and interspersed randomly with oxymethylene groups has been described by Booth and co-workers.[98,99] The polymer

$$\left+\!\!-\!\!(OCH_2CH_2)_m OCH_2\!\!-\!\!\right]_n$$

was prepared from various low-molecular-weight polyethylene glycols (molecular weights 200, 400, and 600) and dichloromethane or dibromomethane as linking agent. Molecular weights as high as 1×10^5 may be obtained, but depend critically on the experimental conditions. The melting points of polymers prepared from PEG 600, PEG 400, and PEG 200 are 26, 14, and $-9°C$, respectively. The latter two are therefore completely amorphous at room temperature. $LiCF_3SO_3$[99]-, and $LiClO_4$[100,101]-based systems of these polymers have been studied over various concentration ranges.

Passiniemi et al.[102] synthesized a random PEO–PPO copolymer by copolymerizing ethylene oxide and propylene oxide in various mole ratios. A polymeric organotin butyl phosphate coordination catalyst was used. Molecular weights of the final polymers varied between 5000 and 53,000. Increasing the PPO content tended to decrease the copolymer T_g and the melting point, whereas complexes with $LiClO_4$ at a constant O:Li ratio of 9:1 showed a general decrease in T_g and ionic conductivity. Przyluski et al.[103,104] synthesized random copolymers of PEO and PPO, polyepichlorohydrin and polystyrene oxide, and carried out a systematic study of these polymers complexed with varying concentrations of NaI, LiI, $LiClO_4$, and $LiBF_4$. X-ray and DSC studies showed the materials to be amorphous. The best conductivities were found for PEO–PPO–$LiBF_4$ films.

6.5. Blends

Blends of a third component with polymer electrolytes can be used to serve one of several purposes. One type of blend is a "gel electrolyte," where a plasticizer (low-molecular-weight solvent) is added to increase ionic conductivity. These electrolytes are not true polymer electrolytes as has been defined here, as ionic transport resembles that of a liquid system and the polymer serves primarily as a support for the conducting matrix. Gel electrolytes based on polymer hosts such as PEO, polystyrene, poly(vinyl chloride), poly(vinyl alcohol), polyacrylonitrile, and poly(vinylidene fluoride), $LiClO_4$ and a high-dielectric-constant solvent have been reported in the literature[105–110] and have been reviewed.[111] Cameron et al.[112] have carried out a quantitative study on blends of liquid polyethers incorporating $LiClO_4$ and propylene carbonate (PC) or tetrahydrofuran (THF) plasticizer. Addition of either plasticizer reduced the viscosity and increased the ionic conductivity of the system. However, the molal conductivity (Λ) over a wide salt concentration range was quite different for the two plasticized systems. The increase was modest on addition of THF, whereas addition of PC produced an abnormally large increase. In addition, the latter system showed a continuous rise in Λ toward low salt concentrations, unlike THF plasticized materials, which showed the more normal maximum in Λ (see Chapter 9). It was proposed that in PC systems, the ionic motion is "decoupled" from that of the polymeric solvent and, because of the higher permittivity of the PC, stronger interactions with the salt may be expected, reducing ion pairing.

Kaplan et al.[113] prepared a macromolecular homolog of propylene carbonate which is a high-T_g glassy material. By addition of a crown ether in stoichiometric ratio with the salt, conductivity was significantly enhanced. As indicated by Armand and Gauthier,[114] the quantities of crown ether are too small to significantly affect the polymer T_g and conductivity may involve the transport of solvated cations or, as suggested by the authors, the crowns are hindered in the polymer matrix and ion transport occurs by exchange with nearest-neighbor crown ether molecules. A 1:1:20 mixture of a crown polymer, $LiCF_3SO_3$, and vinylene carbonate was polymerized and gave similar conductivities.[115]

Kelly et al.[109,110] improved the conductivity of PEO-based polymer electrolytes by adding low-molecular-weight dimethoxy polyethylene glycols, $CH_3(OCH_2CH_2)_n$ OCH_3. The materials behave very similarly to solutions of the salt in the low-molecular-weight ether, showing conductivities of the order of 10^{-4} S cm^{-1}. Mechanical properties were, however, poor. Low-molecular-weight PEO plasticized polyether networks complexed with $LiCF_3SO_3$[116] and the formation of interpenetrating networks of low-molecular-weight PEO, PPO, and PEI supported by a crosslinked polymer such as epoxy, poly(methyl methacrylate), and polyacrylonitrile have been described.[117] Conductivities of the order of 4 to 6×10^{-4} S cm^{-1} were realized for the latter systems but phase separation was found when the support polymer concentration was above 40 wt%.

As the amorphous form of simple high-molecular-weight polymer electrolytes is often susceptible to creep, particularly at elevated temperatures, blends specifically designed to improve mechanical stability have been studied. The earliest reported blend of this type was by Weston and Steele,[118] who incorporated various proportions of α-alumina into $PEO–LiClO_4$. Above 20% by volume, the filler aggregated to form large insulating regions. At 10% filler, the conductivity was not significantly impaired but there was no evidence of creep up to 120°C. The long-term possibility of the filler settling out may make this simple method impractical.

Tsuchida et al.[119] studied systems of $PEO–LiClO_4$ (molecular weight 400–20,000) supported by poly(methyacrylic acid) ($T_g = 128°C$). Despite increased mechanical stability, the conductivities were poor, which may be a consequence of coupling between the polar groups on the polymers,[120] hindering ionic motion. The compatibility of these two polymers may be a result of this interpolymer interaction, but generally miscibility and prevention of phase separation limit the application of this technique for improving mechanical stability. This problem was circumvented by polymerizing styrene after thorough mixing through a $PEO–LiCF_3SO_3$ electrolyte to establish a stiff structural framework at a molecular level.[121,122] Above the percolation limit, polystyrene caused only a small fall in conductivity but significant improvement in mechanical stability at high temperatures. A similar study has been carried out with PEO–MX (MX = NaI, LiI, LiClO$_4$ and LiBF$_4$) blended with poly(methyl methacrylate).[123,124] A room temperature conductivity of 2.4×10^{-4} S cm^{-1} was realized for the polymer blend incorporating LiClO$_4$. Xia et al.[79] copolymerized styrene and methoxypolyethylene glycol methacrylates to improve mechanical stability and reduce crystallinity.

Wnek et al.[42] studied the use of electric fields to modulate the morphology of a 90:10 polystyrene–PEO blend. Homogeneously distributed PEO microspheres exhibited pearl chaining along the electric field lines, and fusion into semicontinuous ribbons can then occur. Exploitation of this effect to produce new polymer electrolytes is possible. Dalal and Crist[125] have described a technique of rolling PEO–salt films to produce an oriented lamellar structure. Anisotropic conductivity parallel and perpendicular to the fibrillar direction was reported. Novel electrolytes have been described that are formed by incorporating a liquid polyether–salt solution into a microporous polyethylene membrane.[126,127] The liquid is immobilized in the pores if wetting properties and pore size and shape are optimized. Such membranes have exhibited conductivities in excess of 10^{-4} S cm^{-1}.

A number of interesting studies on mixed-phase electrolytes have been carried out. These are mixtures of polymer electrolyte (based on PEO) and a crystalline ionic conductor. The crystalline materials have high ionic conductivity and are cation conductors but lack the flexibility of polymer electrolytes; it may therefore be possible to combine the advantageous properties of the two phases. Stevens and

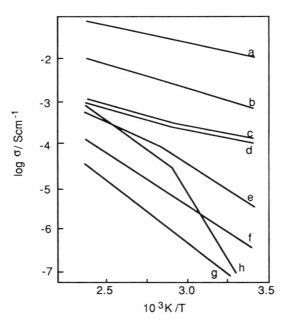

Figure 6.3. Ionic conductivity as a function of temperature for Li$_3$N + P(EO)$_{12}$LiCF$_3$SO$_3$ mixed-phase electrolytes. (**a**) Hydrogen-doped Li$_3$N, (**b**) pure Li$_3$N, (**c**) 5 vol% polymer phase, (**d**) 8%, (**e**) 16%, (**f**) 32%, (**g**) 64%, (**h**) 100%. From Ref. 129.

Mellander[128] studied the electrical conductivity of PEO–MAg_4I_5 (M = Li, K, Rb) compounds. Above the percolation limit for the salt, conductivities increased to give room temperature values of 2×10^{-3} S cm^{-1} for potassium and rubidium compounds and 1×10^{-4} S cm^{-1} for the lithium conductor. The complexes are more complicated than a simple two-phase mixture as AgI and MI were found to be present. The stability of these compounds was poor, with conductivity falling by approximately 60% over 230 days. Skaarup et al.[129] mixed large grains (50–100 μm) of Li_3N with 5 to 20 vol% P(EO)$_{12}$LiCF$_3$SO$_3$. Figure 6.3 shows the conductivities of these mixtures. Optimum conductivity, about three orders of magnitude greater than that of the polymer electrolyte alone, was achieved at 5 to 10% polymer. At higher polymer content, the conductivity decreases but the temperature dependence shows a single Arrhenius-type behavior and an activation energy much lower than that of the pure polymer electrolyte. This indicated variation in the ion transport mechanism between the two systems, with the possibility of the polymer adapting a different structure in the channels between the crystalline grains. Skaarup et al. later investigated a sulfide glass ($1.2Li_2S$–$1.6LiI$–B_2S_3) mixed with different polymer materials.[130] Room temperature ionic conductivities were approximately 1000 times higher than those of the PEO-based polymer electrolytes; however, insulating polyethylene bound systems were shown to give the best conductivities, indicating that the polymer only serves as a binder to keep the grains of the solid phase in contact. Wieczorek and co-workers[131–134] have reported a number of studies of PEO–NaI mixed with the conducting ceramic NASICON ($Na_{3.2}Zr_2Si_{2.2}P_{0.8}O_{12}$) and inert θ-$Al_2O_3$ and SiO_2. Table 6.1 summarizes the properties of the ceramic powders. Hydrophilic SiO_2 (and Al_2O_3) showed a stronger interaction with the polymer than the hydrophobic material, and composites formed with the hydrophilic material showed a consequent reduction in polymer crystallinity and a higher conductivity. Croce et al.[135] have carried out similar studies using β″-Al_2O_3 powder with regular spherical shape and variable grain size finely dispersed in PEO–NaI. The mechanical stability is greatly improved with little effect on the conductivity. Grain size, phase boundary resistances, phase composition, and T_g are all contributing factors to the conductivity of the composite polymer electrolytes and make the analysis of the ion transport very complex.

Table 6.1. Properties of Ceramic Powders Mixed with Polymer Electrolytes

Ceramic Powder	Grain Size (μm)	Surface Area (m² g^{-1})	Affinity to Solvent
NASICON	5	3	—
θ-Al_2O_3	<2	Average area	Hydrophilic
	2–4	700	
	4–7		
	>7		
β-$Al_2O_3^a$	13		
SiO_2	30	103	Hydrophilic
SiO_2	30	24	Hydrophobic

^{a}From Ref. 135.

6.6. Single-Ion Conductors

In a polymer electrolyte, both cation and anion are free to move and, in fact, Cheradame[136] has demonstrated that the main contribution to current transport is by the anion. This is discussed further in Chapter 9. To obtain a transference number of unity, the anion or cation may be immobilized by using a very large counterion or forming a polyelectrolyte where one ionic charge is chemically bonded to the polymer backbone. Two examples of the former are

$$
\begin{array}{cc}
\underset{\substack{\mid \\ \text{—(CH}_2\text{—C)}_n}}{\overset{CH_3}{}} & \underset{\substack{\mid \\ \text{—(CH}_2\text{—C)}_n}}{\overset{CH_3}{}} \qquad \overset{O}{\underset{\parallel}{}} \\
O \qquad OCH_2CH_2SO_3^-Li^+ & O \qquad OCH_2CH_2OC(CF_2)_3COO^-Li^+
\end{array}
$$

poly(2-sulfoethylmethacrylate)   poly(2(4-carboxylhexafluorobutanoloxy)
(PSEM)   ethyl methacrylate) (PCHFEM)

which have been described by Ward and co-workers.[137] These polyelectrolytes were blended with PEO. The conductivities of PCHFEM–PEO films were an order of magnitude higher than those of analogous PSEM–PEO films but were lower than that of a PEO–LiCF$_3$SO$_3$ electrolyte. In polyelectrolytes, unless a solvent is present, tight ion pairing is always likely to occur and mobility of the charge carriers is significantly reduced. Lithium ion complexes of poly(vinyl alcohol) have been reported where conductivities were found to radically improve above a particular ionic group concentration.[138] Presumably the higher concentration facilitated ion hopping between sites. Hardy and Shriver[139] studied the sodium salt of poly(styrene sulfonate) plasticized with PEG. In poly(diallyldimethylammonium chloride) (PDDAC)

$$
\underset{\substack{CH_3 \qquad CH_3}}{\overset{\displaystyle —(CH_2 \qquad CH_2)_n}{\underset{\overset{+}{N}}{\bigvee}}} \quad Cl^-
$$

the substituted quaternary nitrogen prevents close approach of the anion and reduces the ion-pairing problem; however, to obtain reasonable conductivities in light of the material's high glass transition temperature, plasticization with PEG is required.[139,140] Skotheim and co-workers[141] have described the synthesis of a polymer electrolyte based on a nylon-1 backbone, ethylene oxide side chains, and bonded dibutyl phenolate anions:

$$-(N—C)_x \text{———} (N—C)_y$$

[Structure diagram with benzene rings, t-Bu groups, O⁻ M⁺, followed by chain:]

O
|
CH_2
|
CH_2
$)_6$
O
|
CH_3

The tendency for ion pairing appears to be reduced as the conductivity a sodium ion-containing compound reached approximately 2×10^{-6} S cm^{-1} at 50°C.

PDDAC is an example of a class of polymers termed "ionenes" (an abbreviation for ionic amines)[142] that have the general formula

$$+ (R_1—\overset{\overset{\displaystyle R_3}{|}}{\underset{\underset{\displaystyle R_3}{|}}{N^+}}—R_2—\overset{\overset{\displaystyle R_3}{|}}{\underset{\underset{\displaystyle R_3}{|}}{N^+}} —)$$
$$X^- \qquad\qquad\qquad X^-$$

A wide variety of aliphatic, aromatic, and heterocyclic ionenes can be synthesized and range in properties from highly crystalline through polymeric to glassy solids. Their properties are determined by the nature of the polymer chains and by ionic interactions that depend on the density of charge on the backbone. PPO–ionene copolymers have been described earlier where the ionene imparts mechanical strength to the conducting system. An example of an ionene as a component of a polyelectrolyte complex of a polycation and polyanion is given later. Meyer and co-workers[143–146] have recently carried out extensive studies on the properties of ionenes as possible candidates for polymer electrolytes. Table 6.2 lists the thermal properties of a large variety of ionenes. Materials have yet to be optimized with respect to ion mobility, that is, segmental motion in main or side chains and minimization of ion pairing. These materials have been reviewed in detail recently[146] and are not discussed further here.

Polyelectrolytes based on flexible backbone polymers have been synthesized by a number of groups. Networks containing phosphate charged groups or the lithium salt group —$OSbCl_5^-Li^+$ have been described by Cheradame and co-workers.[62,147] Shriver[148] has reported the synthesis and properties of polyphosphazene sulfonates

Table 6.2. Thermoanalytical Data for Various Ionenes

Chemical Structure of Backbone	Counterion X⁻	T_g	Melting Point (°C)	T_{decomp} (°C)
—(CH₂)₆—N⁺(Me)₂—	CH₃—C₆H₄—SO₃⁻	—	249	280
—(CH₂)₆—N⁺(Me)₂—	CF₃SO₃⁻	—	206	300
—(CH₂)₆—N⁺(Me)₂—	BF₄⁻	—	276	335
—(CH₂)₆—N⁺(Me)₂—	PF₆⁻	—	230	230
—(CH₂)₆—N⁺(Me)₂—	CH₃—C₆H₄—SO₃⁻	29	119	—
—(CH₂)₁₀—N⁺(Me)₂—	CH₃—C₆H₄—SO₃⁻	42	99	300
—(CH₂)₁₀—N⁺(Me)₂—	CF₃SO₃⁻	—	118	260
—(CH₂)₁₀—N⁺(Me)₂—	BF₄⁻	38	127	300
—(CH₂)₁₀—N⁺(Me)₂—	PF₆⁻	—	226	250
—(CH₂)₆/₁₀—N⁺(Me)₂—	BF₄⁻	60	—	—
—(CH₂)₁₀—N⁺(Me)₂—	SbF₆⁻	—	283	283
—(CH₂)₁₀—N⁺(Me)₂—	ZnBr₄²⁻	60	278	278
—(CH₂)₁₀—N⁺(Me)₂—	BF₄⁻	-22	—	—
—(CH₂)₄—O—(CH₂)₄—N⁺(Me)₂—	BF₄⁻	-17	—	—
—(CH₂)₂—O—(CH₂)₂—O—(CH₂)₂—N⁺(Et₂)—	BF₄⁻	-30	122	—
—(CH₂)₂₀—N⁺(Me)₂—	BF₄⁻	—	74	—
—(CH₂)₂₀—N⁺(Me)₂—	ZnBr₄²⁻	99	232	280
—CH₂—C₆H₄—OCO—C₆H₄—OCO—C₆H₄—CH₂—N⁺(Me)₂—	BF₄⁻	156	—	—
—CH₂—C₆H₄—SO₂—C₆H₄—CH₂—N⁺(Me)₂—	ClO₄⁻	193	—	225
—CH₂—C₆H₄—O—C₆H₄—CH₂—N⁺(Me)₂—	PF₆⁻		—	—

$$-\!\!\!\!-\!\!\big[O(CH_2CH_2O)_yCH_3\big]_x\!\!-\!\!\!\!-$$

$$[N\!=\!P]_n$$

$$(OCH_2CH_2SO_3{}^-Na^+)_{2-x}$$

$y = 2$ and $x = 1.54, 1.75, 1.8$

$y = 7.22$ and $x = 1.8$

The elastomeric properties were a function of the sulfonate content and at best the conductivities were two orders of magnitude lower than that of a MEEP–NaSO$_3$SO$_3$ analog. This was attributed to extensive ion pairing. The behavior of this polymer was compared with that of a polyether network structure incorporating a tetra-(alkoxy)aluminate counterion.[149] The equivalent conductivity of this latter system was relatively constant with concentration, whereas the former showed a decrease as the ionizable group concentration rose. The T_g increased with concentration for the network, whereas the comb-type polymer showed little change below a sulfonate group concentration of 10%. These differences may be the result of the different structure type or an effect of ion pairing.

Comb-type lithium and sodium polyelectrolytes based on half esters of maleic and anhydride–styrene copolymers with polyethylene glycol pendant groups have been described[150]:

$$(CH\!-\!CH\!-\!CH_2\!-\!CH)_n$$

$$O\!=\!C \quad C\!=\!O$$

$$-\!O$$

$$M^+ \qquad O(CH_2CH_2O)_n\!-\!H$$

The conductivity values varied with PEG molecular weight and cation from 1×10^{-9} to 1×10^{-5} S cm^{-1}; the best values were found for molecular weights 200 to 600. The low conductivities may well result from steric hindrance: a larger EO group unable to move sufficiently and a shorter chain less able to coordinate the neighboring ion effectively. Rietman and Kaplan[151] have reported the properties of styrene–maleic anhydride and ethylene–maleic anhydride backbone comb-type polymer electrolytes. Again, the best conductivities achieved after optimizing the PEO chain length were of the order of 5×10^{-7} S cm^{-1}.

Kobayashi et al.[152,153] have described the preparation of poly[oligo-(oxyethylene)methacrylate-*co*-(alkali metal methacrylate)]s, prepared by radical polymerization of the alkali metal methacrylate and oligo(oxyethylene) methacrylate, CH$_3$C(CH$_2$)COO(CH$_2$CH$_2$O)$_7$CH$_3$. The ionic conductivity of Li and K salts is 10^{-6} S cm^{-1} at 80°C. Toyota et al.[154] considered a highly polar polymeric medium rather than a low-molecular-weight solvent to bring about complete salt dissociation. LiClO$_4$ was dispersed in a polycation–polyanion matrix of the form

$$\left[\begin{array}{c} \mathrm{-CH-CH_2CH-CH_2-} \\ | \quad\quad | \\ \mathrm{COO^-} \quad \mathrm{COO^-} \\ \mathrm{Li^+} \quad\quad \mathrm{ClO_4^-} \\ \mathrm{CH_3} \quad\quad \mathrm{CH_3} \\ | \quad\quad\quad | \\ \mathrm{-N^+ \!\!-\!(CH_2)_3\!N^+\!\!-\!(CH_2)_6} \\ | \quad\quad\quad | \\ \mathrm{CH_3} \quad\quad \mathrm{CH_3} \\ \mathrm{ClO_4^-} \quad \mathrm{Li^+} \end{array} \right]_n$$

but only modest conductivities were obtained, possibly because of the strong association of Li^+ ions with carboxylate groups and interchain interactions imparting rigidity on the matrix. Tsuchida and co-workers[155] formed compatible blends of low-molecular-weight polyethers with the perfluoro polyelectrolytes Nafion and Flemion:

$$(C_2F_4)_x \!-\! (C_2F_3)_y \qquad\qquad (C_2F_4)_x \!-\! (C_2F_3)_y$$
$$| \qquad\qquad\qquad\qquad\qquad |$$
$$O \qquad\qquad\qquad\qquad\qquad O$$
$$| \qquad\qquad\qquad\qquad\qquad |$$
$$C_3F_6 \qquad\qquad\qquad\qquad\quad C_3F_6$$
$$|_z \qquad\qquad\qquad\qquad\qquad |_z$$
$$OC_2F_4SO_3^-Li^+ \qquad\qquad COO^-Li^+$$

$$\text{Nafion} \qquad\qquad\qquad\qquad \text{Flemion}$$

These polymers are known to form microphase separated structures containing cylindrical anionic domains[156] into which the polyether can be introduced. With component compositions optimized, conductivities of the order of 10^{-5} S cm^{-1} can be realized at room temperature.

6.7. Salts

All the methods of maximizing the conductivity so far discussed have been based on modifications to the polymer host. The electrolyte properties may also be varied by changing the salt. The influence of the nature and concentration of the incorporated salt on the electrolyte is complex as can be seen from the phase diagrams in Chapter 4. Criteria for the formation of polymer electrolytes were discussed in Chapter 3. Essentially, polymer electrolytes form when the salt consists of a polarizing cation and a large anion of delocalized charge to minimize the lattice energy. The salt affects the conductivity through crystalline complex formation, intramolecular crosslinking of the polymer chains, and the degree of salt dissociation (the number of charge carriers). In particular the anion has a major influence on phase composi-

Table 6.3. Lithium Salts with Possible Applications in Polymer Electrolytes

COMMON LITHIUM SALTS
LiCl, LiBr, LiI, LiSCN, LiClO$_4$, LiNO$_3$

LITHIUM SALTS WITH LARGE ANIONS
Monovalent anions
  LiBϕ_4, Li(C$_5$H$_{11}$—C=C)$_4$B, Li(C$_4$H$_9$—C=C)$_4$B, Li(C$_6$H$_5$—((CH$_2$)$_3$—C=C)$_4$B
  R—COO—CH$_2$—CH$_2$—SO$_3$Li (R = non-solvating polymer chain)
Divalent anions
  Li$_2$B$_{10}$Cl$_{10}$, Li$_2$B$_{12}$Cl$_{12}$, Li$_2$B$_{12}$H$_{12}$

FLUORO COMPOUNDS
Monovalent anions
  LiCF$_3$SO$_3$, LiC$_4$F$_9$SO$_3$, LiC$_6$F$_{13}$SO$_3$, LiC$_8$F$_{17}$SO$_3$, LiCF$_3$CO$_2$, LiN(CF$_3$CO$_2$)$_2$, LiN(CF$_3$SO$_2$)$_2$,
    LiN(CH$_3$SO$_2$)$_2$, LiAsF$_6$, LiBF$_4$
Divalent anions
  LiOOC(CF$_2$)$_3$COOLi, LiSO$_3$(CF$_2$)$_3$SO$_3$Li

tion and conductivity. Table 6.3 lists most of the lithium salts that have been used to form polymer electrolytes. Armand and Elkadiri[157] have proposed criteria for solubility, conductivity, and redox stability requirements for polymer electrolytes in rechargeable batteries that restrict this list. So far, LiClO$_4$ and LiCF$_3$SO$_3$ are the only common salts that fulfill the electrochemical stability criteria, although new perfluorosulfonimide salts, for example, (CF$_3$SO$_2$-N-SO$_2$CF$_3$)$^-$Li$^+$, have been successfully tested in high-power cells.[158]

Symmetry of the anion or flexibility of its chain can sometimes introduce a plasticizing effect that lowers the T_g and thus enhances conductivity.[159] Anions of the form

have been synthesized. Moderate enhancement in conductivity was reported and was attributed to the low polarity of the anionic group. Although very flexible, the groups were unable to modify the chain dynamics in the vicinity of the cations. Less polar PPO showed a more marked improvement than the PEO complexes. Imide anions of the type $(X-N-X)^-$, where X is an electron-withdrawing group, have been studied.[158,159] Such molecules are mechanically flexible, have large delocalization of charge, and exhibit weaker interaction between alkali metals and nitrogen than with polyether oxygen while having a large electronegativity. Armand et al.[160] described the synthesis and conducting properties of the perfluorosulfonimide systems. At room temperature, systems with O:Li = 3–10 appear to be amorphous. Conductivities of approximately 5×10^{-5} S cm^{-1} at 25°C, almost independent of composition, were attained, which is a marked improvement in PEO–LiCF$_3$SO$_3$ complexes.

References

1. C. Berthier, W. Gorecki, M. Minier, M. B. Armand, J. M. Chabagno, and P. Rigaud, *Solid State Ionics* **11** (1983), 91.

2. M. B. Armand, J. M. Chabagno, and M. Duclot, in *Fast Ion Transport in Solids* (P. Vashista, J. N. Mundy, and G. K. Shenoy, Eds.), Elsevier/North-Holland, New York (1979), p. 131.

3. C. R. A. Catlow, A. V. Chadwick, G. N. Greaves, L. M. Moroney, and M. R. Worboys, *Solid State Ionics* **9/10** (1983), 1107.

4. P. G. Bruce, F. Krok, J. Evans, and C. A. Vincent, *Br. Polym. J.* **20** (1988), 193.

5. R. D. Armstrong and M. D. Clarke, *Electrochim. Acta* **29** (1984), 443.

6. J. Morgan, A. McLennaghan, and R. A. Pethrick, *Eur. Polym. J.* **25** (1989), 1087.

7. D. F. Shriver, B. L. Papke, M. A. Ratner, R. Dupon, T. Wong, and M. Brodwin, *Solid State Ionics* **5** (1981), 83.

8. R. Dupon, B. L. Papke, M. A. Ratner, and D. F. Shriver, *J. Electrochem. Soc.* **131** (1984), 586.

9. M. Watanabe, M. Rikukawa, K. Sanui, N. Ogata, H. Kato, T. Kobayashi, and Z. Ohtaki, *Macromolecules* **17** (1984), 2902.

10. M. Watanabe, M. Rikukawa, K. Sanui, and N. Ogata, *Macromolecules* **19** (1986), 188.

11. M. Watanabe, M. Togo, K. Sanui, N. Ogata, T. Kobayashi, and Z. Ohtaki, *Macromolecules* **17** (1984), 2908.

12. P. M. Blonsky, D. F. Shriver, P. Austin, and H. R. Allcock, *J. Am. Chem. Soc.* **106** (1984), 6854.

13. D. F. Shriver, S. Clancy, P. M. Blonsky, and L. C. Hardy, in *Proceedings of the 6th Risø International Symposium on Metallurgy and Materials Science* (F. W. Poulsen, N. Hessel Andersen, K. Clausen, S. Skaarup and O. T. Sørensen, Eds.), Risø National Laboratory, Roskilde (1984).

14. G. Nazri, D. M. MacArthur, and J. F. Ogara, *Chem. Mater.* **1** (1989), 370.

15. S. G. Greenbaum, K. J. Adamic, Y. S. Pak, M. C. Wintersgill, and J. J. Fontanella, *Solid State Ionics* **28–30** (1988), 1042.

16. G. A. Nazri, D. M. MacArthur, and J. F. Ogara, *Polym. Prepr.* **30**, No. 1 (1989), 430.

17. M. M. Lerner, L. J. Lyons, J. S. Tonge, and D. F. Shriver, *Polym. Prepr.* **30**, No. 1 (1989), 435.

18. P. M. Blonsky, D. F. Shriver, P. Austin, and H. R. Allcock, *Solid State Ionics* **18/19** (1989), 258.

19. H. R. Allcock, P. E. Austin, T. X. Neenan, J. T. Sisko, P. M. Blonsky, and D. F. Shriver, *Macromolecules* **19** (1986), 1508.

20. J. S. Tonge and D. F. Shriver, *J. Electrochem. Soc.* **134** (1987), 270.

21. J. L. Bennet, A. A. Dembek, H. A. Allcock, B. J. Heyen, and D. F. Shriver, *Polym. Prepr.* **30**, No. 1 (1989), 437.

22. H. R. Allcock, R. J. Fitzpatrick, M. Gebura, and S. Kwon, *Polym. Prepr.* **28** (1987), 321.

23. G. Nazri and S. P. Meibuhr, in *Materials and Processes for Lithium Batteries* (K. M. Abraham and B. B. Owens, Eds.), Electrochem. Soc., Pennington, N. J., 89–4 (1989), p. 322.

24. K. Nagaoka, H. Naruse, I. Shinohara, and M. Watanabe, *J. Polym. Sci. Polym. Lett. Ed.* **22** (1984), 659.

25. P. G. Hall, G. R. Davies, J. E. McIntyre, I. M. Ward, D. J. Bannister, and K. M. F. Le Brocq, *Polym. Commun.* **27** (1986), 98.

26. D. Fish, I. M. Khan, and J. Smid, *Makromol. Chem., Rapid Commun.* **7** (1986), 115.

27. R. S. Spindler and D. F. Shriver, *J. Am. Chem. Soc.* **110** (1988), 3036.

28. D. Fish, I. M. Khan, E. Wu, and J. Smid, *Br. Polym. J.* **20** (1988), 281.

29. I. M. Khan, Y. Yuan, D. Fish, E. Wu, and J. Smid, *Macromolecules* **21** (1989), 2684.

30. D. J. H. Ballard, P. Cheshire, T. S. Mann, and J. E. Przeworski, *Macromolecules* **23** (1990), 1256.

31. G. Zhou, I. M. Khan, and J. Smid, *Polym. Prepr.* **30,** No. 1 (1989), 416.

32. G. Zhou, I. M. Khan, and J. Smid, *Second International Symposium on Polymer Electrolytes (ISPE-2), Siena, June 14–16* (1989), Extended Abstracts, p. 46.

33. T. F. Yeh, H. Liu, Y. Okamoto, H. S. Lee, and T. A. Skotheim, in *Second International Symposium on Polymer Electrolytes* (B. Scrosati, Ed.), Elsevier, London (1990), p. 83.

34. A. Bouridah, F. Dalard, D. Deroo, H. Cheradame, and J. F. Le Nest, *Solid State Ionics* **15** (1985), 233.

35. L. Lestel, S. Boileau, and H. Cheradame, *Polym. Prepr.* **30,** No. 1 (1989), 133.

36. H. Cheradame, A. Killis, L. Lestel, and S. Boileau, *Polym. Prepr.* **30,** No. 1 (1989), 420.

37. L. Lestel, S. Boileau, and H. Cheradame, in *Second International Symposium on Polymer Electrolytes* (B. Scrosati, Ed.), Elsevier, London (1990), p. 83.

38. L. Lestel, S. Boileau, and H. Cheradame, *Polymer* **31** (1990), 1154.

39. M. Watanabe, S. Nagano, K. Sanui, and N. Ogata, *J. Power Sources* **20** (1987), 327.

40. K. J. Adamic, S. G. Greenbaum, M. C. Wintersgill, and J. J. Fontanella, *J. Appl. Phys.* **60** (1986), 1342.

41. S. B. Fang, P. Zhang, and Y. Y. Jiang, *Polym. Bull.* **18** (1988), 81.

42. G. E. Wnek, K. Gault, J. Serpico, C. Y. Yang, G. Venugopal, and S. Krause, in *Second International Symposium on Polymer Electrolytes* (B. Scrosati, Ed.), Elsevier, London (1990), p. 73.

43. Y. Uchimoto, Z. Ogumi, and Z. Takehara, *Solid State Ionics* **35** (1989), p. 417.

44. Z. Ogumi, Y. Uchimoto, Z. Takehara, and F. R. Foulkes, *J. Electrochem. Soc.* **137** (1990), 29.

45. Y. Uchimoto, Z. Ogumi, and Z. Takehara, *Solid State Ionics* **40/41** (1990), p. 624.

46. Y. Uchimoto, Z. Ogumi, Z. Takehara, and F. R. Foulkes, *J. Electrochem. Soc.* **137** (1990), 35.

47. J. M. G. Cowie, A. C. S. Martin, and A. M. Firth, *Br. Polym. J.* **20** (1988), 247.

48. S. Pantaloni, S. Passerini, F. Croce, B. Scrosati, A. Roggero, and M. Andrei, *Electrochim. Acta* **34** (1989), 635.

49. M. Andrei, L. Marchese, and A. Roggero, in *Second International Symposium on Polymer Electrolytes* (B. Scrosati, Ed.), Elsevier, London (1990), p. 107.

50. A. Killis, J. F. Le Nest, and H. Cheradame, *Makromol. Chem. Rapid Commun.* **1** (1980), 595.

51. H. Cheradame, J. L. Souquet, and J. M. Latour, *Mater. Res. Bull.* **15** (1980), 1173.

52. A. Killis, J. F. Le Nest, A. Gandini, and H. Cheradame, *P. Polym. Sci. Polym. Phys. Ed.* **19** (1981), 1073.

53. H. Cheradame, A. Gandini, A. Killis, and J. F. Le Nest, *J. Power Sources* **9** (1983), 389.

54. A. Killis, J. F. Le Nest, A. Gandini, H. Cheradame, and J. P. Cohen-Addad, *Solid State Ionics* **14** (1984), 231.

55. A. Killis, J. F. Le Nest, A. Gandini, and H. Cheradame, *Macromolecules* **17** (1984), 63.

56. D. Andre, J. F. Le Nest, and H. Cheradame, *Eur. Polym. J.* **17** (1981), 57.

57. A. Killis, J. F. Le Nest, H. Cheradame, and A. Gandini, *Makromol. Chem.* **83** (1982), 2835.

58. A. Killis, J. F. Le Nest, H. Cheradame, and A. Gandini, *Makromol. Chem.* **83** (1982), 1037.

59. M. Leveque, J. F. Le Nest, H. Cheradame, and A. Gandini, *Makromol. Chem. Rapid Commun.* **4** (1983), 497.

60. M. Leveque, J. F. Le Nest, A. Gandini, and H. Cheradame, *J. Power Sources* **14** (1985), 27.

61. D. Deroo, A. Bouridah, F. Dalard, H. Cheradame, and J. F. Le Nest, *Solid State Ionics* **15** (1985), 233.

62. H. Cheradame and J. F. Le Nest, in *Polymer Electrolyte Reviews—1* (J. R. MacCallum and C. A. Vincent, Eds.), Elsevier, London (1987), p. 103.

63. J. F. Le Nest, A. Gandini, and H. Cheradame, *Macromolecules* **21** (1988), 1117.

64. J. F. Le Nest, A. Gandini, and H. Cheradame, *Br. Polym. J.* **20** (1988), 253.

65. M. Watanabe, K. Sanui, N. Ogata, and F. Inoue, *Polym. J.* **16** (1984), 711.

66. M. Watanabe, K. Sanui, N. Ogata, T. Kobayashi, and Z. Ohtaki, *J. Appl. Phys.* **57** (1985), 123.

67. M. Watanabe, K. Sanui, N. Ogata, F. Inoue, T. Kobayashi, and Z. Ohtaki, *Polym. J.* **17** (1985), 549.

68. M. Watanabe, S. Nagano, K. Sanui, and N. Ogata, *Solid State Ionics* **18/19** (1986), 338.

69. M. Watanabe, A. Suzoki, T. Santo, K. Sanui, and N. Ogata, *Macromolecules* **19** (1986), 1921.

70. M. Watanabe, K. Sanui, an 1 N. Ogata, *Macromolecules* **19** (1986), 815.

71. M. Watanabe and N. Ogata, in *Polymer Electrolyte Reviews—1* (J. R. MacCallum and C. A. Vincent, Eds.), Elsevier, London (1987), p. 39.

72. J. R. M. Giles and M. P. Greenhall, *Polym. Commun.* **27** (1987), 360.

73. J. R. M. Giles, *Solid State Ionics* **24** (1987), 155.

74. J. R. MacCallum, M. J. Smith, and C. A. Vincent, *Solid State Ionics* **11** (1981), 307.

75. E. Kronfli, K. V. Lovell, A. Hooper, and R. J. Neat, *Br. Polym. J.* **20** (1988), 275.

76. A. Le Mehaute, G. Crepy, G. Marcellin, T. Hamaide, and A. Guyot, *Polym. Bull.* **14** (1985), 233.

77. A. Le Mehaute, T. Hamaide, G. Crepy, and G. Marcellin, European Patent Application 8619049 (1983).

78. K. Ishikawa, T. Sugihara, Y. Ohshima, T. Kato, and A. Imai, *Solid State Ionics* **40/41** (1990), p. 612.

79. D. W. Xia, D. Soltz, and J. Smid, *Solid State Ionics* **14** (1984), 221.

80. D. W. Xia and J. Smid, *J. Polym. Sci. Polym. Lett. Ed.* **22** (1984), 617.

81. D. J. Bannister, G. R. Davies, I. M. Ward, and J. E. McIntyre, *Polymer* **25** (1984), 1600.

82. N. Kobayashi, M. Uchiyama, K. Shigehara, and E. Tsuchida, *J. Phys. Chem.* **89** (1985), 987.

83. N. Kobayashi, M. Uchiyama, and E. Tsuchida, *Solid State Ionics* **17** (1985), 307.

84. J. M. G. Cowie and A. C. S. Martin, *Polymer Commun.* **26** (1985), 298.

85. J. M. G. Cowie and R. Ferguson, *J. Polym. Sci. Polym. Phys. Ed.* **23** (1985), 2181.

86. J. M. G. Cowie, In *Polymer Electrolyte Reviews—1* (J. R. MacCallum and C. A. Vincent, Eds.), Elsevier, London (1987), p. 69.

87. J. M. G. Cowie and A. C. S. Martin, *Polymer* **28** (1987), 627.

88. J. M. G. Cowie, R. Ferguson, and A. C. S. Martin, *Polym. Commun.* **28** (1987), 130.

89. Y. Yamaguchi, S. Aoki, M. Watanabe, K. Sanui, and N. Ogata, *Solid State Ionics* **40/41** (1990), 628.

90. M. Watanabe, S. Oohashi, K. Sanui, N. Ogata, T. Kobayashi, and Z. Ohataki, *Macromolecules* **18** (1985), 1945.

91. M. Watanabe, K. Nagaoka, M. Kanba, and I. Shinohara, *Polym. J.* **14** (1982), 877.

92. C. Robitaille and J. Prud'homme, *Macromolecules* **16** (1983), 665.

93. F. M. Gray, J. R. MacCallum, C. A. Vincent, and J. R. M. Giles, *Macromolecules* **21** (1988), 392.

94. J. R. M. Giles, F. M. Gray, J. R. MacCallum, and C. A. Vincent, *Polymer* **28** (1987), 1977.

95. J. S. Foos and S. M. Erker, *J. Electrochem. Soc.* **134** (1987), 1724.

96. G. Goulart, S. Sylla, J. V. Sanchez, and M. Armand, in *Second International Symposium on Polymer Electrolytes* (B. Scrosati, Ed.), Elsevier, London (1990), p. 99.

97. J. R. M. Giles, C. Booth, and R. H. Mobbs, Transport–Structure Relations in Fast Ion and Mixed Conductors, in *Proceedings, 6th Risø International Symposium on Metallurgy and Materials Science* (F. W. Poulsen, N. Hassel Andersen, K. Clausen, S. Skaarup, and O. T. Sørensen, Eds.), Risø National Lab., Roskilde (1985), p. 329.

98. J. R. Craven, R. H. Mobbs, C. Booth, and J. R. M. Giles, *Makromol. Chem. Rapid Commun.* **7** (1986), 81.

99. C. V. Nicholas, D. J. Wilson, C. Booth, and J. R. M. Giles, *Br. Polym. J.* **20** (1988), 289.

100. E. Linden and J. R. Owen, *Solid State Ionics* **28–30** (1988), 994.

101. F. M. Gray, *Solid State Ionics* **40/41** (1990), 637.

102. P. Passiniemi, S. Takkumäki, J. Kankare, and M. Syrjämä, *Solid State Ionics* **28–30** (1988), 1001.

103. J. Przyluski, W. Wieczorek, Z. Florjanczyk, and W. Krawiec, *Second International Symposium on Polymer Electrolytes (ISPE-2), Siena, June 14–16* (1989), Extended Abstracts, p. 56.

104. Z. Florjanczyk, W. Krawiec, W. Wieczorek, and J. Przyluski, to be published.

105. E. Tsuchida, H. Ohno, and K. Tsunemi, *Electrochim. Acta* **28** (1983), 591.

106. K. Tsunemi, H. Ohno, and E. Tsuchida, *Electrochim. Acta* **28** (1983), 833.

107. M. Watanabe, M. Kanba, H. Matsuda, K. Tsunemi, K. Mizoguchi, E. Tsuchida, and I. Shinohara, *Macromol. Chem. Rapid Commun.* **2** (1981), 741.

108. M. Watanabe, M. Kanba, K. Nagaoka, and I. Shinohara, *J. Polym. Sci. Polym. Phys. Ed.* **21** (1983), 939.

109. I. Kelly, J. R. Owen, and B. C. H. Steele, *J. Electroanal. Chem. Interfacial Electrochem.* **168** (1984), 467.

110. I. Kelly, J. R. Owen, and B. C. H. Steele, *J. Power Sources* **14** (1985), 13.

111. F. M. Gray, in *Polymer Electrolyte Reviews—1* (J. R. MacCallum and C. A. Vincent, Eds.), Elsevier, London (1987), p. 139.

112. G. G. Cameron, M. D. Ingram, and K. Sarmouk, *Eur. Polym. J.* **26** (1990), 197.

113. M. L. Kaplan, E. A. Rietman, R. J. Cava, L. K. Holt, and E. A. Chandross, *Solid State Ionics* **25** (1987), 37.

114. M. Armand and M. Gauthier, in *High Conductivity Solid Ionic Conductors. Recent Trends and Applications* (T. Takahashi, Ed.), World Science, Singapore (1989), p. 114.

115. M. Kaplan, E. A. Rietman, and R. J. Cava, *Polymer* **30** (1989), 504.

116. J. Tsuchiya, in *Solid State Ionics,* (G. Nazri, R. A. Huggins, and D. F. Shriver, Eds.), **135,** Materials Research Society, Pittsburgh (1989), p. 357.

117. B. J. Bauer, C. K. Chiang, and G. T. Davis, U.S. Patent 4654279 (1987).

118. J. E. Weston and B. C. H. Steele, *Solid State Ionics* **7** (1982), 75.

119. E. Tsuchida, H. Ohno, K. Tsunemi, and N. Kobayashi, *Solid State Ionics* **11** (1983), 227.

120. T. Ikawa, K. Abe, K. Honda, and E. Tsuchida, *J. Polym. Sci. Polym. Chem. Ed.* **13** (1975), 1505.

121. F. M. Gray, J. R. MacCallum, and C. A. Vincent, *Solid State Ionics* **18/19** (1986), 252.

122. F. M. Gray, J. R. MacCallum, and C. A. Vincent, British Patent Application 8619049.

123. K. Such, Z. Florjanezyk, W. Wieczorek, and J. Przyluski, in *Second International Symposium on Polymer Electrolytes* (B. Scrosati, Ed.), Elsevier, London (1990), p. 119.

124. Z. Florjanezyk, K. Such, W. Wieczorek, and J. Przyluski, 7th International Conference on Solid State Ionics, Hakone, November 5–11 (1989), Extended Abstracts, p. 187.

125. V. F. Dalal and B. Crist, *Polym. Prep* **30**, No. 1 (1989), 426.

126. T. Itoh, K. Koseki, K. Mukaida, K. Kohno, Q. J. Cao, and O. Yamamoto, *Solid State Ionics* **40/41** (1990), 620.

127. K. Koseki, S. Saeki, T. Itoh, C. Zao, and O. Yamaguchi, in *Second International Symposium on Polymer Electrolytes* (B. Scrosati, Ed.), Elsevier, London (1990), p. 197.

128. J. R. Stevens and B. E. Mellander, *Solid State Ionics* **21** (1986), 203.

129. S. Skaarup, K. West, and B. Zachau-Christiansen, *Solid State Ionics* **28–30** (1988), 975.

130. S. Skaarup, K. West, P. M. Julian, and D. M. Thomas, *Solid State Ionics* **40/41** (1990), p. 1021.

131. J. Plocharski and W. Wieczorek, *Solid State Ionics* **28–30** (1988), 979.

132. J. Przyluski and W. Wieczorek, in *Solid State Ionic Devices* (B. V. R. Chowdari and S. Radhakrishna, Eds.), World Science, Singapore (1988), p. 475.

133. W. Wieczorek, K. Such, J. Plocharski, and J. Przyluski, in *Second International Symposium on Polymer Electrolytes* (B. Scrosati, Ed.), Elsevier, London (1990), p. 339.

134. W. Wieczorek, K. Such, H. Wycislik, and J. Plocharski, *Solid State Ionics* **36** (1989), 255.

135. F. Croce, F. Bonimo, S. Panero, and B. Scrosati, *Phil. Mag. B* **59** (1989), 161.

136. H. Cheradame, J. F. LeNest, A. Gandini, and M. Leveque, *J. Power Sources* **14** (1985), 27.

137. D. J. Bannister, G. R. Davies, I. M. Ward, and J. E. McIntyre, *Polymer* **25** (1984), 1291.

138. G. L. Bao, W. Q. Yun, S. B. Fang, and Y. Y. Jiang, *Polym. Bull.* **18** (1987), 143.

139. L. C. Hardy and D. F. Shriver, *J. Am. Chem. Soc.* **107** (1985), 3823.

140. L. C. Hardy and D. F. Shriver, *Macromolecules* **17** (1984), 975.

141. H. Liu, Y. Okamoto, T. Skotheim, Y. S. Pak, S. G. Greenbaum, and K. J. Adamic, in *Solid State Ionics,* (G. Nazri, R. A. Huggins, and D. F. Shriver, Eds.), **135** Materials Research Society, Pittsburgh (1989), p. 337.

142. A. Rembaum, *J. Macromol. Sci. Chem.* **3** (1969), 87.

143. L. Dominquez and W. H. Meyer, *Solid State Ionics* **28–30** (1988), 941.

144. L. Dominquez, W. H. Meyer, and G. Wegner, *Makromol. Chem. Rapid Commun.* **8** (1987), 151.

145. L. Dominguez, V. Enkelmann, W. H. Meyer, and G. Wegner, *Polymer* **30** (1989), 2030.

146. W. H. Meyer and L. Dominquez, in *Polymer Electrolyte Reviews—2* (J. R. MacCallum and C. A. Vincent, Eds.), Elsevier, London (1989), p. 191.

147. J. F. LeNest, A. Gandini, H. Cheradame, and J. P. Cohen-Addad, *Polym. Commun.* **28** (1987), 302.

148. D. F. Shriver, *Macromolecules* **21** (1988), 2299.

149. K. E. Doan, S. Ganapathiappan, K. Chen, M. A. Ratner, and D. F. Shriver, in *Solid State Ionics* (G. Nazri, R. A. Huggins, and D. F. Shriver, Eds.), **135,** Materials Research Society, Pittsburgh (1989), p. 343.

150. J. Przyluski, W. Krawiec, T. Listos, W. Wieczorek, and Z. Florjanezyk, *7th International Conference on Solid State Ionics,* Hakone, November 5–11 (1989), Extended Abstracts, p. 188.

151. E. A. Rietman and M. L. Kaplan, *J. Polym. Sci. Polym. Lett. Ed.* **28** (1990), 187.

152. N. Kobayashi, T. Hamada, H. Ohno, and E. Tsuchida, *Polym. J.* **18** (1986), 661.

153. E. Tsuchida, N. Kobayashi, and H. Ohno, *Macromolecules* **21** (1988), 96.

154. S. Toyota, T. Nogami, and H. Mikawa, *Solid State Ionics* **13** (1984), 243.

155. K. Shigehara, N. Kobayashi, and E. Tsuchida, *Solid State Ionics* **14** (1984), 85.

156. I. M. Hodge and A. Eisenberg, *Macromolecules* **11** (1978), 289.

157. M. B. Armand and F. Elkadiri, in *Proceedings, Symposium on Lithium Batteries*, San Diego, 1986 (A. N. Dey, Ed.), 87-1, The Electrochem. Soc., Pennington, N.J. (1987), p. 502.

158. M. Gauthier, A. Bélanger, B. Kapfer, G. Vassort, and M. Armand, in *Polymer Electrolyte Reviews—2* (J. R. MacCallum and C. A. Vincent, Eds.), Elsevier, London (1989), p. 285.

159. M. F. Gauthier, M. Armand, and D. Muller, *Electroresponsive Molecules and Polymeric Systems*, Vol. 1 (T. A. Skotheim, Ed.), Dekker, New York (1988).

160. M. B. Armand, W. Gorecki, and R. Andreani, in *Second International Symposium on Polymer Electrolytes* (B. Scrosati, Ed.), Elsevier, London (1990), p. 187.

CHAPTER 7

Further Developments in Polymer Electrolyte Materials

7.1. Proton Conductors

Extensive research into polymer electrolytes in general has been carried out since the late 1970s; however, studies on proton-conducting polymeric materials have been reported only since the latter part of the 1980s.

Ethers form hydrogen bonds but they also tend to be attacked by strong aqueous acids:

$$—O— + H^+ \rightleftharpoons —O— \xrightarrow[\ ]{H^+ \quad H_2O} —OH \quad HO—$$

Armand and co-workers[1-3] showed that in the absence of moisture, degradation is suppressed and well-defined complexes such as poly(ethylene oxide) (PEO) with phosphoric acid, $P(EO)_n H_3 PO_4$, can be formed.

7.1.1. Polyethers

A phase diagram for $P(EO)_n H_3 PO_4$ has been constructed and is shown in Figure 7.1. Two eutectics melting close to room temperature are formed. Conductivity isotherms have been superimposed,[4] and it is observed that optimum conductivity is found for a value of $n = 0.5$, which coincides with a eutectic point. Figure 7.2 shows the temperature variation of the conductivity for a number of acid concentrations and it is seen that reasonable room temperature values (4×10^{-5} S cm^{-1}) can be achieved. Quasi-elastic neutron scattering (QENS) and nuclear magnetic resonance (NMR) have been used[3,5,6] to investigate the dynamics of PEO–$H_3 PO_4$ electrolytes. The diffusion coefficient of the proton measured by pulsed field gradient NMR and the diffusion coefficient of chain segments determined by QENS indicate that their mobilities are closely related. Comparison of the proton diffusion coefficient with conductivity data suggests that only a small percentage of the protons participate in the charge transport at any one time.

7.1.2. Polyamides

Polyamides such as polyvinylpyrolidone (PVP) and polyacrylamide (PAAM)[5-7] form complexes with inorganic acids similar to those the polyethers form. They

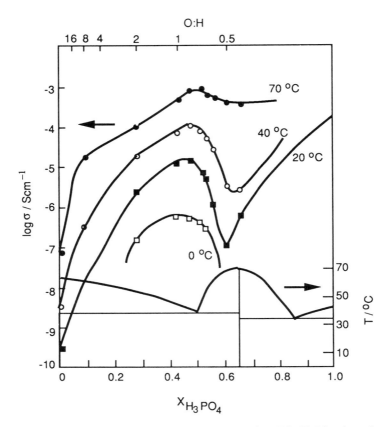

Figure 7.1. Phase diagram and conductivity isotherms for PEO–H$_3$PO$_4$ electrolytes. From Ref. 4.

possess a basicity comparable to that of water and the ethers and can be protonated at either the oxygen or nitrogen:

$$-(CH_2-CH)- \quad \xrightarrow{H^+} \quad -(CH_2-CH)- \quad \rightleftharpoons \quad -(CH_2-CH)- \quad \rightleftharpoons \quad -(CH_2-CH)-$$

PVP

Because of the polarity of polyamides, the glass transition temperatures are high: T_g = 200°C for PVP; however, addition of acid plasticizes the systems. For a PVP:H$_3$PO$_4$ ratio of 2:1, the T_g is approximately 200°C, at 1:1, it is approximately

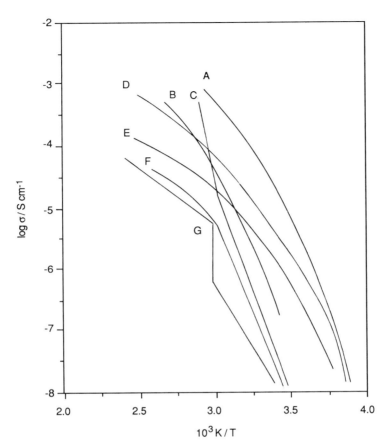

Figure 7.2. Conductivity dependence on temperature for P(EO)$_x$H$_3$PO$_4$ at different salt concentrations. O:H = (**A**) 0.8, (**B**) 0.6, (**C**) 0.4, (**D**) 1, (**E**) 2, (**F**) 8, and (**G**) 16.

100°C, and at 1:2 it falls to approximately 0°C.[8] The conductivity does not, however, follow the T_g. At a ratio of 2:1 $\sigma = 4 \times 10^{-9}$ S cm^{-1} at 130°C, peaks at 3×10^{-5} S cm^{-1} at 20°C for a 2:3 ratio, and falls to 10^{-6} S cm^{-1} for a 1:2 ratio at room temperature.[8] The temperature variation of the conductivity is shown in Figure 7.3. In the case of PAAM–H$_2$SO$_4$ systems, it has been demonstrated through infrared (IR) spectroscopy that only one proton participates in the protonation of the polymer and the anion HSO$_4^-$ is formed. Although protonation of the —CONH$_2$ group gives essentially —CONH$_3^+$, there is evidence that modifications to the electronic distribution of the C—N and C=O bonds also occurs. The concentration variation of the conductivity rises sharply with increasing acid to a constant value. It is not certain whether intrachain proton exchange between several resonance forms may contribute to the very high room temperature conductivity of 10^{-2} S cm^{-1} (Figure 7.4) observed for the PAAM–H$_2$SO$_4$ 1:1.5 electrolyte.

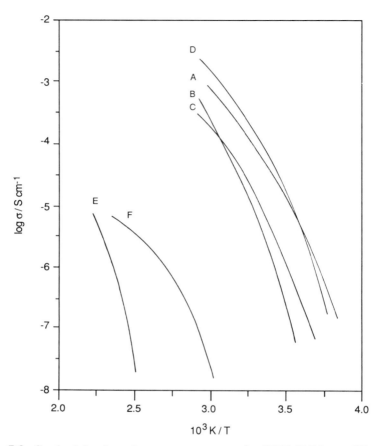

Figure 7.3. Conductivity dependence on temperature for $P(VP)_xH_3PO_4$ at different salt concentrations. Polymer repeat unit:acid ratio = (**A**) 2:3, (**B**) 1:2, (**C**) 3:4, (**D**) 3:5, (**E**) 2:1, and (**F**) 1:1.

7.1.3. Poly(acrylic acid)

Poly(acrylic acid) (PAA) mixed with NH_4HSO_4 (which is a proton conductor itself) can exhibit room temperature conductivity better than 10^{-5} S cm^{-1}, but it is very dependent on the preparation technique.[8] Infrared spectroscopy revealed that no real complexation with the polymer occurred and it was assumed that conductivity came about both by NH_4^+ jumps and by proton migration along HSO_4^- hydrogen-bonded chains.

7.1.4. Polyamines

Both linear and branched polyethylene imine (LPEI, BPEI)[7–9] coordinate with inorganic acids to form proton-conducting electrolytes. At low protonation levels

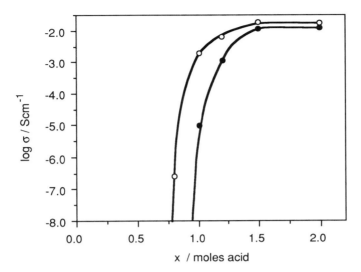

Figure 7.4. Concentration dependence of the conductivity for $PAAM_xH_2SO_4$ at 20°C (○) and 100°C (●). From Ref. 5.

(acid-to-monomer unit ratio < 0.15) the materials are amorphous elastomers; at higher acid content the former form crystalline species and the latter are vitreous materials. The conductivities for LPEI and BPEI are 10^{-8} S cm^{-1} and 8.5×10^{-3} S cm^{-1}, respectively, for a concentration of 0.35 mol of H_2SO_4 per monomer unit. For LPEI–H_3PO_4 systems the conductivity is found to be two orders of magnitude higher than that for LPEI–H_2SO_4. The reverse situation is observed for the branched polymer complexes. The concentration dependence of the conductivities of BPEI systems is shown in Figure 7.5. At low acid levels, IR spectroscopy shows protonation of the polymer and SO_4^{2-} anions,[8] and it has been proposed that the PEI–acid conductivity mechanism involves proton hopping between coordinating sites on the polymer, whereas at higher doping levels, proton motion on the anion lattice, for example, $HSO_4^- \leftrightarrow SO_4^{2-}$ or $H_2PO_4^- \leftrightarrow HOP_4^{2-}$ may occur.[7,8] Similar concentration–temperature dependencies are observed for PEO–H_3PO_4[4] systems; however, the situation must be rather more complex than suggested here as the nature of the polymer, the T_g, and morphology clearly have contributing roles in the conductivity mechanism. T_g (Figure 7.5) increases from $-47°C$ for pure BPEI to 37°C for an acid-to-monomer ratio of approximately 0.35. This is due to an increase in NH . . . . O hydrogen bonds as the polymer becomes protonated. The decrease in T_g at increased acid levels coincides with the increase in conductivity. It was proposed[8] that new OH . . . . O interactions between HSO_4^- and SO_4^{2-} compete with NH . . . . O hydrogen bonds and greater flexibility is introduced into the matrix. Watanabe et al.[10] have studied proton conduction in crystalline and amorphous PEI hydrates as a function of water content. It was noted that the conductivity in the crystal hydrates of the linear form followed an Arrhenius temperature dependence with an activation energy that was largely independent of composition.

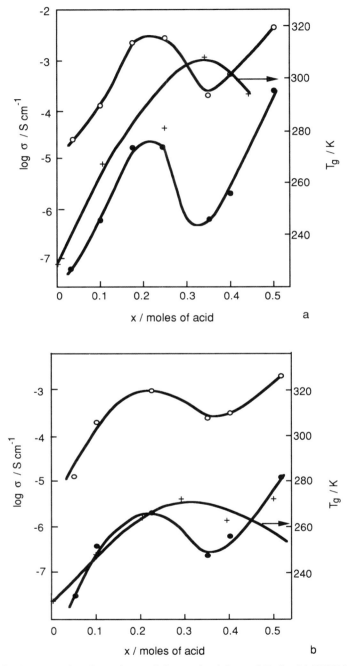

Figure 7.5. Concentration dependence of the conductivity and T_g for (a) $BPEI_xH_2SO_4$ and (b) $BPEI_xH_3PO_4$ at 300 K (○) and 373 K (●). From Ref. 5.

7.1.5. Poly(vinyl alcohol)

Properties of poly(vinyl alcohol) (PVA)–H_3PO_4 and PVA–H_2SO_4 proton conductors have been reported by Singh et al.[11] Conductivities show rather unusual temperature and concentration dependence. For the former material, two plateaus are observed, one of which was attributed to the T_g of PVA. The conductivity of the phosphoric acid systems was reported to increase rapidly with increasing acid content and reached a maximum at approximately 40 wt% acid. It was noted that dry samples could not be formed at these high acid concentrations. Polak et al.[12] have also studied electrical properties of PVA and phosphoric acid complexes. Again, these materials were reported to contain water and, as pointed out by Armand and Gauthier,[4] it is probable that H_3PO_4 causes the dehydration of PVA to form polyacetylene.

In summary, PAAM–H_2SO_4 and BPEI–H_2SO_4 systems exhibit by far the best conductivities of the proton conductors where conductivities of the order of 10^{-2} S cm^{-1} at room temperature are possible. At present, there are insufficient data to explain the observed conductivity differences between polymer matrices and, particularly, the extremely high values for the aforementioned systems.

7.2. Ormocers

To prepare materials with novel properties, it is often necessary to develop composites of common structures to produce a final material that combines the desirable properties of each component. It was for the purpose of synthesizing ceramics that would meet a range of specifications, for example, in purity or homogeneity, which was not possible with readily available materials, that the sol–gel process for producing inorganic networks was developed.[13,14] One of the major advantages of this technique is that network formation of the inorganic polymer is carried out at low temperatures in organic or aqueous solution. This subsequently led to an expanded range of materials through the building of organic components into the inorganic network. Different mixed organic–inorganic hybrid materials have been developed over the last decade[14] and have become known as ceramers or ormosils (ORganically MOdified SILicates) or, less specifically, as ormocers (ORganically MOdified CERamics).

7.2.1. The Sol–Gel Process

Over the past 25 years, a systematic development of ceramic materials using what is known as the sol–gel process has taken place, with particular interest in silica-containing systems. Many features of this technique have made it particularly attractive:

1. During the network growth, several different polycondensation reactions may occur. The properties of the final material can be tailored by inhibiting or promoting particular pathways.

2. Homogeneous distributions of low-concentration components, for example, salts for ionic conductivity, can be achieved.
3. Ceramics and glasses can be formed at low temperatures. As a consequence, organic components can be incorporated into the network.

The synthetic route to an inorganic-based network takes place in solution, producing first a colloidal phase, but, with the continuing reaction, a solid gel is finally formed. The basic processing steps may be summarized as

precursor → hydrolysis → reactive monomer → condensation → sol
gelation → gel

Most work involving sol–gel syntheses has been carried out using alkoxides as precursors, as they can easily form reactive monomers that are in general soluble in common solvents. The precursor need not be an alkoxide: soluble salts, oxides, chelates, hydroxides, and, where the particle sizes are small enough to obtain a good homogeneous distribution of components, colloids may be used. This factor, however, tends to restrict the use of the latter materials as precursors. For solutions, the reactive monomers are generated by hydrolysis, for example:

$$AlCl_3 + H_2O \rightarrow Cl_2AlOH + HCl$$
$$Si(OR)_4 + H_2O \rightarrow (RO)_3SiOH + HOR$$

For colloids, the reaction is activated by destabilization through surface charge control. An additional advantage of the solution route is that the rate of hydrolysis can be controlled, which is important in determining the properties of the final gel.[15]

The processing steps involve polycondensation reactions, the two most important being

$$\equiv M\text{—}OH + X\text{—}M\equiv \rightarrow \equiv M\text{—}O\text{—}M\equiv + H\text{—}X$$
$$\equiv M\text{—}OH + HO\text{—}M\equiv \rightarrow \equiv M\text{—}O\text{—}M\equiv + H_2O$$

Where, for example, M = Al, Si, Ti, and X = OH, OR, halogen, OCOR, NR_2. The types of materials, based on alkoxide precursors, that can be synthesized by this method and a summarized representation of the sol–gel processing of ceramics and glasses and ormocers are shown schematically in Figures 7.6 and 7.7.[16]

In general, many factors influence the outcome of the sol–gel process: solvent, temperature, ligands, order of compound addition, hydrolysis rate, and, for ormocers, steric influence of the organic groups and lead to a very large variety of polymeric structures. The chemistry is complex, and to process particular material structures is an art in itself. A more comprehensive discussion can be found in reviews by Schmidt.[13,14]

7.2.2. Ormolytes

Ormolytes are the electrolytic forms of ormocers. The proton donor is normally added during the last steps of the reaction.[17] Armand, Poinsignon, and others have

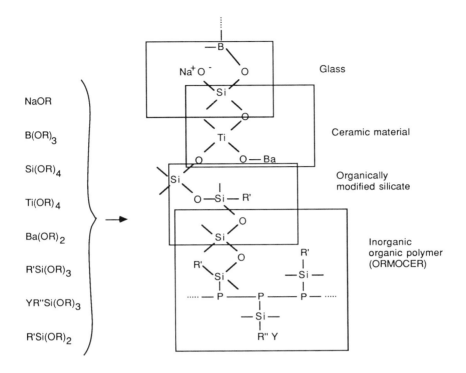

P = polymerisable ligand
R = CH$_3$, C$_2$H$_5$
R' = (CH$_2$)$_3$NH$_2$, C$_6$H$_5$
R" = ionic ligand
Y = ions

Figure 7.6. Types of materials, based on alkoxide precursors, that can be synthesized by the sol–gel process.

described the synthesis and properties of a family of organically modified silicates referred to as aminocils. The electrolytes are nonstoichiometric and have the general formula[17–21]

$$SiO_{3/2}R—(HX)_x$$

where HX represents the acid and R is an amino group, typically $(CH_2)_3NH_3$, $—(CH_2)_4NH_2$, $—(CH_2)_3NH(CH_2)_2NH$, $—(CH_2)_3NH(CH_2)_2NH(CH_2)_2NH_2$, and $—(CH_2)_3N(CH_3)_2$. x refers to the ratio HX:Si.[18,19] These are synthesized by mixing together organotrialkoxysolanes $(R'O)_3SiR$ where R' is $—CH_3$ or $—C_2H_5$, an acid HX, and water $(H_2O:Si = 6)$ which gives the reactions

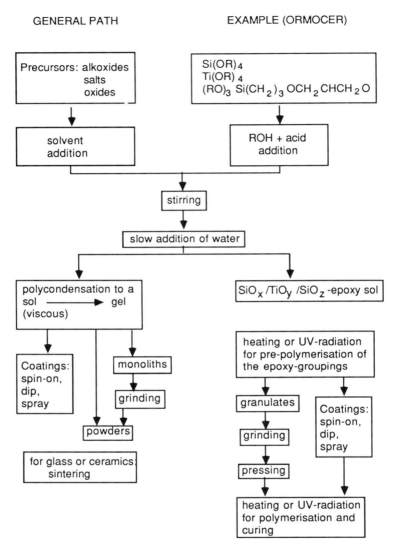

Figure 7.7. Sol–gel processing of glasses, ceramics, and a specific example of an ORMOCER.

$$(R'O)_3SiR \xrightarrow[-nR'OH]{+nH_2O} (R'O)_{3-n}(OH)_nSiR \qquad \text{hydrolysis}$$

$$\equiv SiOH + HOSi\equiv \xrightarrow[-H_2O]{} \equiv SiOSi\equiv \qquad \text{polycondensation}$$

Properties of aminosils complexed with HCF_3SO_3, HCH_3SO_3, HCl, HNO_3, HCH_3COO, and H_3PO_4 have been reported. Of the acids, CH_3COOH and H_3PO_4

complexes differ markedly from the others. Acetic acid is the weakest and proton transfer is incomplete. Films are heterogeneous. Phosphoric acid complexes are more rigid than the others, suggesting polyfunctional crosslinking or incorporation of the phosphate group into the silica network:

$$H_3PO_4 + 3\equiv SiOH \longrightarrow \equiv Si-O-\underset{\underset{\underset{\overset{|||}{Si}}{|}}{\overset{\overset{O}{||}}{\underset{|}{O}}}{P}-O-Si\equiv + 3H_2O$$

The materials are found to form transparent, hard films that are nonporous, stable in the atmosphere, and thermally stable up to 180°C. The temperature variation of the conductivity shows non-Arrhenius behavior and room temperature values of the order of 3×10^{-5} S cm^{-1} may be achieved. Longer alkylamine chains appear to be detrimental to conductivity.[18]

7.3. Multivalent Cation-Based Polymer Electrolytes

As for the proton conductors, since the mid-1980s interest has started to grow in polymer electrolytes containing salts based on divalent cations and, even more recently, trivalent cationic systems. The understanding of these materials is currently comparable to that of lithium-based systems of the early 1980s, and they have only recently begun to be studied by a systematic method of approach. The range of multivalent cations that can in principle be incorporated into suitable polymer hosts will lead to a far wider diversity of properties than has been the case for monovalent systems. This, in turn, will give rise to possible new applications, some of which will require ionic motion, whereas others will not. For example, as magnesium and calcium are divalent, anodes based on these metals have effectively twice the capacity of lithium and may therefore lead to less expensive electrochemical cells. Also, certain lanthanides are known to have interesting phosphorescent properties and are currently used in oxide glass matrices. It may be possible to develop these materials as low-cost flexible phosphors. Many other properties have the potential for exploitation, particularly mixed-valence, for example, Co^{2+}/Co^{3+}, cationic matrices, which could, in principle, be developed as amorphous electronic conductors. The potential for these materials is promising but a basic understanding of the physical and chemical properties has yet to be realized.

7.3.1. Preparation of Materials

The majority of studies on multivalent cation-containing polymer electrolytes are based on a PEO host polymer. One study has been reported for a number of divalent and trivalent salts in poly(methoxyethoxy ethoxy)phosphazene (MEEP).[22] As with alkali metal–PEO electrolytes, the particular microstructure in any multivalent electrolyte is a complex function of composition and thermal history and presents difficulties when comparing properties of different electrolytes. Film preparation is normally by solvent casting as described in Chapter 4. It is, however, apparent that reproducibility of results is highly dependent on the approach adopted for film formation and is *critically* dependent on levels of moisture contamination. Farrington and Linford[23] have described a method leading to films of reproducible conductivity. Moisture-free electrolyte films may be produced by heating under vacuum; however, this may in fact affect the final morphology of the material.[24,25] Also, if moisture is present during preparation, for example, as a result of forming samples in the open atmosphere, and is removed afterward, rather than preparing under scrupulously dry conditions, it is found that variations in the local structure of the polymer film are introduced. The effects of hydration/dehydration on the conductivity of PEO–NiBr$_2$ films can be seen in Figure 7.8. The process appears to change the temperature dependence of the conductivity, enhancing it at higher temperatures. Indeed, the Ni^{2+} ion is found to be mobile after hydration/ dehydration,[26] while effectively immobile when prepared under strictly anhydrous conditions. As discussed in Chapter 4, the casting solvent may also affect the morphology of the final film. A number of solvents such as acetonitrile[27] or solvent mixes such

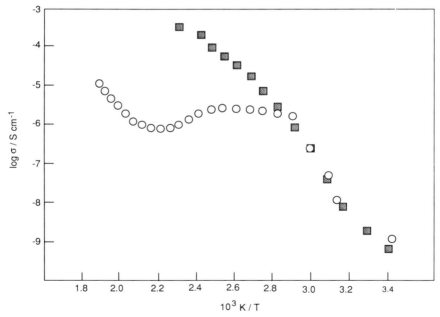

Figure 7.8. Conductivity of P(EO)$_8$NiBr$_2$ before (○) and after (■) deliberate hydration and dehydration of the sample. From Ref. 26. Reprinted by permission of the publisher, The Electrochemical Society, Inc.

as dimethylsulfoxide/acetonitrile[29] or ethanol/acetonitrile[26,29,30] have been used for PEO–divalent cation salt film formation.

As many divalent and trivalent salts are neither obtained nor easily handled in the anhydrous form, an effective method of preparing anhydrous multivalent polymer electrolytes was proposed by Gray.[31] This involves refluxing an acetonitrile solution of the hydrated salt, allowing the solvent to condense through a bed of molecular sieves. A solution of the polymer is subsequently added and the solvent slowly removed. In this way potentially dangerous anhydrous perchlorates are never isolated. It is found, however, that some cations such as mercury[32] can catalyze the decomposition of the solvent:

$$(CH_3CN)_xHg^{2+} \xrightarrow[slow]{H_2O} CH_3-C{=}N...Hg^{2+}(CH_3CN)_{x-1}$$

with $\underset{H}{\overset{O}{\diagdown}}\underset{H}{\diagup}$ group

$$\xrightarrow{fast} CH_3-C{=}N...Hg^{2+}(CH_3CN)_{x-1}$$

with OH and H groups

$$\rightleftharpoons CH_3-\underset{\|}{\overset{O}{C}}-N{\overset{H}{\underset{H}{\diagup}}}Hg^{2+}(CH_3CN)_{x-1}$$

Mercury perchlorate could be dehydrated by equilibrating the salt solution with ion-exchanged molecular sieves.[33]

7.3.2. Physical Properties

An essential prerequisite for significant cation mobility is a high lability of the cation–polymer bond. Thus, while strong cation–polymer bonds are necessary for polymer electrolyte formation, labile bonds are necessary for cation mobility. The rate of exchange of solvent coordinated to a cation is affected by many factors: cation radius and charge are important, but for transition metal ions, the dominant factor is often the change in d-orbital energy on going from a ground-state solvation structure (e.g., octahedral coordination) to a transition state (e.g., square pyramidal or bipyramidal coordination). Eigen[34] showed that Hg^{2+} has the fastest exchange rate with water of all the divalent cations (Figure 7.9). Its exchange rate ($\sim10^9$ s^{-1}) is comparable to that of lithium ($\sim5 \times 10^8$ s^{-1}). Because of the similar donicities of the oxygen in water and ethers, exchange rates in ethers can be expected to follow a similar trend. It was predicted[27] that Hg^{2+} might therefore be mobile in PEO. Ca^{2+} with a water exchange rate of approximately 4×10^8 s^{-1} is also expected to be mobile. On the other hand, Mg^{2+} has a water exchange rate of approximately

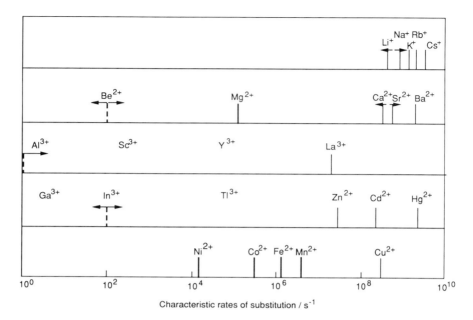

Figure 7.9. Characteristic rate constants for H_2O substitution in the inner coordination sphere of metal ions. Reprinted with permission from Ref. 34, *Pure and Applied Chemistry* **6**, M. Eigen, Copyright 1968, Pergamon Press Plc.

10^5 s^{-1} and should be much less mobile. The trivalent cations of the lanthanide group have water exchange rates of the order of 10^7 to 10^8 s^{-1} and may on this premise be again expected to exhibit mobility in an anhydrous polymer electrolyte. Conductivity and transference number measurements have shown that although Hg^{2+} is indeed mobile[27] and Mg^{2+} immobile,[35] Ca^{2+} and La^{3+} are also immobile.[32] On the basis of the hard–soft acid–base principle the hardness of divalent cations follows the trend[36]

$$Mg^{2+}, Ca^{2+} > Ni^{2+}, Cu^{2+}, Zn^{2+}, Co^{2+}, Pb^{2+} > Cd^{2+}, Hg^{2+}$$

Therefore, in this instance, one might expect both Mg^{2+} and Ca^{2+} to be immobile and Hg^{2+} to be mobile. La^{3+} is one of the hardest of the lanthanide ions but, in addition, the charge on the La^{3+} ion will undoubtedly contribute further to its lack of mobility. That Hg^{2+} is a soft cation and shows very high lability may account for the very reasonable conductivities observed for anhydrous PEO–$Hg(ClO_4)_2$ systems.[27] The conductivities of lithium, lanthanum, and mercury perchlorate electrolytes where O:M = 8:1 are compared in Figure 7.10.

Other factors such as electronegativity and ionic radius may also affect the mobility of the ions. More important, the mobile species are likely to be associated (as in 1:1 electrolytes) and the nature of these species and degree of association will depend on the anion, cation, and salt concentration. The majority of studies on

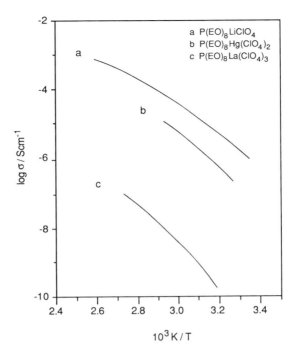

Figure 7.10. Total conductivity of PEO–metal perchlorate electrolytes with an EO:salt ratio of 8:1.

divalent cation polymer electrolytes have involved divalent halides as the salt. This series of salts has an associated problem in that results may not be directly compara-ble because there is considerable variation in the character of the anion–cation bond. For example, the chlorides of zinc, cadmium, and mercury show a sharp transition from ionic to covalent character. Likewise, the tendency of, for example, zinc halides, to form complexes decreases on going from the chloride to the iodide. Cadmium halides in general form auto complex ions such as CdX^+, CdX_3^-, and $Cd\,X_4^{2-}$ in concentrated solutions and the iodides dissociate in solution as

$$2CdI_2 \rightleftharpoons Cd^{2+} + CdI_4^{2-}$$

to give totally anomalous transference number data. Mercury halides are too covalent to allow the free Hg^{2+} ion to form in solution. Thus, $HgCl_2$ dissociates only slightly to give

$$HgCl_2 \rightleftharpoons HgCl^+ + Cl^-$$

These factors are important considerations when interpreting data as the mobile species may vary from electrolyte to electrolyte. In addition to the species, the morphology of the polymer electrolyte is important when determining transference number data. Many of the electrolytes containing divalent or trivalent cations include high-melting complexes melting at 180°C and upward.[29,32] In many instances, measurements may have been made on heterogeneous systems.

7.3.2.1. Alkaline Earth Metal Cations

PEO complexes of $MgCl_2$,[24,28] $Mg(ClO_4)_2$,[37-39] $Mg(SCN)_2$,[39] and $Mg(CF_3SO_3)_2$[40] have been reported in the literature. $P(EO)_4MgCl_2$ electrolytes were found to contain crystalline PEO, indicating that the crystalline complex has a composition in which $EO:Mg^{2+}$ is less than 4.[35] For a 16:1 complex, at 100°C, conductivities were found to be 1.5×10^{-5} S cm^{-1} for $PEO-MgCl_2$, 1.6×10^{-4} S cm^{-1} for $PEO-Mg(CF_3SO_3)_2$, 2×10^{-4} S cm^{-1} for $PEO-Mg(ClO_4)_2$, and 4×10^{-6} S cm^{-1} for $PEO-Mg(SCN)_2$. Apart from the thiocyanate complex, conductivities are comparable to those of $PEO-LiCF_3SO_3$ complexes above 80°C. As the Mg^{2+} transference number was found to be essentially zero it can be concluded that these are anion conductors.

Transference number measurements have been carried out on $PEO-Ca(ClO_4)_2$[32] and $PEO-CaI_2$[23] and it is found that the cation is effectively immobile. X-ray absorption fine-structure (EXAFS) studies on $PEO-CaI_2$ systems[37,40] have shown the Ca^{2+} ion to be trapped within a cage of 10 ether oxygens, which may account for its immobility. Conductivity data have been reported for a number of calcium salt–PEO electrolytes where the anion is SCN^-,[39] $CF_3SO_3^-$,[22,38] ClO_4^-,[32,39] or I^-.[37] Some discrepancy is noted between studies of the perchlorate systems but this may be caused by moisture as only one study was carried out under anhydrous conditions.[32] The conductivities are shown in Figure 7.11. As observed in a number of divalent cation-containing electrolytes, unusual behavior is observed at high salt content. This effect is not understood. As with magnesium systems, the thiocanate

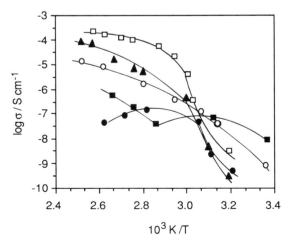

Figure 7.11. Temperature dependence of the conductivity of $P(EO)_xCa(ClO_4)_2$. $x = 50$ ($\square$), 20 ($\blacktriangle$), 12 ($\bigcirc$), 8 ($\blacksquare$), and 4 ($\bullet$).

complexes show conductivities lower by a factor of approximately 10 than the other electrolytes.

X-ray diffraction and differential scanning calorimetry studies of PEO–$Ca(ClO_4)_2$ films indicated the presence of two eutectics, one with a composition between 20:1 and 12:1 and a second with composition near 8:1. Both crystalline complexes are high melting (>200°C). Conductivity data have been reported for MEEP–$Sr(CF_3SO_3)_2$[22] and PEO–$Sr(ClO_4)_2$[39] electrolytes which are similar to PEO–$Mg(ClO_4)_2$ systems. The latter electrolytes were prepared from the hydrated salt and were dried at room temperature only.

7.3.2.2. Zinc(II)

Electrolytes containing ZnX_2, where X= Cl, Br, ClO_4, and CF_3SO_3,[23,37,38,41–44] have been described for a wide range of salt concentrations. Highest conductivities were found for $P(EO)_{16}ZnI_2$ (3.6×10^{-4} S cm^{-1}) and $P(EO)_{20}Zn(CF_3SO_3)_2$ (3.1×10^{-4} S cm^{-1}) at 140°C. $P(EO)_8ZnI_2$ was found to be amorphous at room temperature.[29] Transference number measurements made by the ac and dc polarization techniques (see Chapter 9) showed the cation to be essentially immobile in anhydrous zinc chloride samples[29] but to have a T_+ value of 0.9 in the hydrated form. Substantial ion clustering in $P(EO)_8ZnI_2$ has been suggested from local structure analysis carried out using EXAFS.[37,45] It has been reported that in $P(EO)_nZnCl_2$ electrolytes, two defined complexes and two eutectics at compositions $n = 30$ and 14 exist.

7.3.2.3. Cadmium(II)

PEO–cadmium halide electrolytes contain a crystalline complex that melts at about 180°C. These electrolytes have to be prepared by mixed solvent casting and their properties are somewhat affected by the solvent ratio. A rather high (0.8) cadmium ion transference number has been obtained for a $P(EO)_8CdBr_2$ electrolyte,[29] but this may be an anomaly resulting from complex species formation. Effects of ion association on transference number measurements are discussed in Chapter 9.

7.3.2.4. Lead(II)

Like the cadmium halides, the cation in lead halide–PEO complexes appears to be mobile. Both Pb^{2+} and Cd^{2+} are more polarizable than Mg^{2+} or Ca^{2+} and should therefore have weaker ion–ether oxygen coordination; however, Saraswat et al.[46] report the transference number for Pb^{2+} in PEO–$Pb(ClO_4)_2$ systems to be near zero. PEO–$PbBr_2$ films show higher conductivities ($\sim 5 \times 10^{-6}$ S cm^{-1}) than PEO–PbI_2 samples ($\sim 10^{-6}$ S cm^{-1}) at 140°C. This compares with a value of approximately 10^{-4} S cm^{-1} for $P(EO)_{40}(ClO_4)_2$ and $P(EO)_{16}PbCl_2$.[28] The values for the halides are considerably lower than those for zinc halide-based electrolytes. As the lead halide–PEO crystalline complex melts below 140°C[29] morphological considerations are not required when comparing these two families of electrolytes.

7.3.2.5. Cobalt(II), Nickel(II), and Manganese(II)

Results have been presented for PEO–CoBr$_2$, PEO–MnBr$_2$, and PEO–NiBr$_2$ electrolytes.[23,29] As described earlier, the NiBr$_2$ system shows unusual behavior in that the transference number of the nickel ion appears to be a function of hydration history.[26,29] PEO–MnBr$_2$ systems have conductivities of the order of 10^{-4} to 10^{-5} S cm^{-1} between 130 and 150°C and are reported to be far less crystalline than other divalent samples at room temperature. P(EO)$_8$MnBr$_2$ electrolytes are pale yellow and fluoresce pink-orange under ultraviolet light. P(EO)$_8$CoBr$_2$ films are reported to be amorphous at room temperature although Arrhenius-type plots indicate the presence of crystalline PEO. Conductivity values above 80°C are better than those found for P(EO)$_8$ZnBr$_2$ systems, reaching a value of approximately 10^{-4} S cm^{-1} at 140°C.[29]

7.3.2.6. Mercury(II)

Anhydrous mercuric perchlorate–PEO systems have been investigated.[27,32,33,47] These systems are interesting in that there is no evidence for the formation of crystalline complexes. Conductivity data are reported earlier and compare favorably with data for PEO–LiClO$_4$ electrolytes. The Hg^{2+} ion is mobile and a cation transference number of 0.25 ± 0.05 at 53°C is given for P(EO)$_{20}$(Hg(ClO$_4$)$_2$.

7.3.2.7. Copper(II)

Of all the divalent systems studied to date, PEO-containing copper salts have revealed the most intriguing properties. A number of studies have been carried out by Abrantes,[44] Scrosati,[48–50] Ferloni,[51,52] Farrington,[53] Gray,[32] and their co-workers. Films that have been solvent cast were reported to be blue,[49,52] whereas hot-pressed materials that had never been in contact with solvent were found to be yellow-green.[32] The former showed absorption in the ultraviolet–visible spectrum at 750 nm, and the latter materials absorbed at 825 nm. This is most likely a solvent effect: where a hexacoordination with a Cu^{2+} ion is expected for small molecules such as CH$_3$CN, this is much less likely for large PEO molecules and the lower coordination results in a wavelength shift. It has also been found that when left in a drybox atmosphere for many months, blue solvent-cast films turn yellow.[54] Heating these electrolytes under vacuum to remove solvent poses a problem in itself as decomposition of the electrolyte results from prolonged heating at temperatures above approximately 110°C.[55]

A phase diagram for PEO–Cu(CF$_3$SO$_3$)$_2$ is shown in Figure 7.12.[50] Like many divalent systems, this gives rise to two eutectic points. The impedance spectra for copper salt–PEO systems indicate that the behavior of cells containing this type of electrolyte is not straightforward. For example, Figure 7.13(a–c) shows impedance plots for P(EO)$_{50}$Cu(CF$_3$SO$_3$)$_2$ between "blocking" electrodes indium tin oxide (ITO), gold, and stainless steel (SS). Films were prepared by hot pressing in a polytetrafluoroethylene die. None of these show classical blocking behavior.[56] Figure

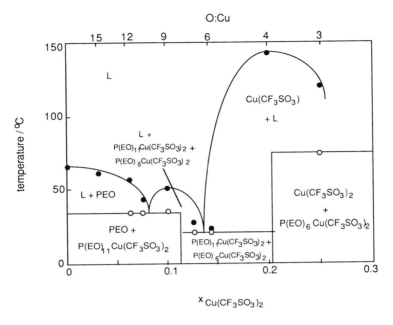

Figure 7.12. Phase diagram for a $PEO-Cu(CF_3SO_3)_2$ system.

7.13d shows the impedance spectrum for the electrolyte under nonblocking conditions. The inset highlights the high-frequency end of the semicircle where an inductive loop is observed. This can sometimes be observed in spectra of blocking cells[56] and appears to be associated with a reduction of Cu^{2+} to Cu^+ at the electrolyte interface. As $CuCF_3SO_3$ is unstable in low-molecular-weight PEO,[56] disproportionation to Cu^0 and Cu^{2+} would be expected. Copper deposits may be observed on the blocking electrode surfaces. The possibility of mixed ionic/electronic conductivity in these systems was suggested by Scrosati and co-workers and further investigations in this area may help to explain many of these observations.

The temperature variation of the conductivity, obtained from the ac spectrum of an $ITO| P(EO)_{50}Cu(CF_3SO_3)_2 |ITO$ cell is shown in Figure 7.14. It is interesting to note that for all PEO–copper salt electrolytes reported,[44,50,56] the activation energy above 60°C is of the order of 8 to 10 kJ mol^{-1}, which is very much lower than those for other polymer electrolyte systems studied. Weston and Steele[57] reported a value of 55 to 60 kJ mol^{-1} for $P(EO)_{10}LiCF_3SO_3$. Many unresolved phenomena in these very interesting systems have yet to be given fuller consideration.

7.3.2.8. Trivalent Cation-Containing Electrolytes

A small number of studies have been carried out on trivalent cation-containing polymer electrolytes and include $PEO-La(ClO_4)_3$,[32,47,58] $PEO-EuCl_3$, PEO–

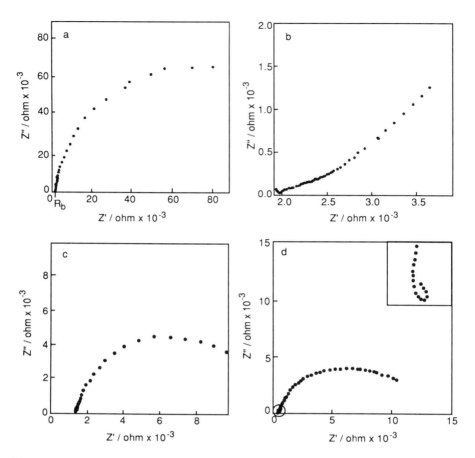

Figure 7.13. Impedance spectra for (a) ITO | P(EO)$_{50}$Cu(CF$_3$SO$_3$)$_2$ | ITO, (b) Au | P(EO)$_{50}$ Cu(CF$_3$SO$_3$)$_2$ | Au, (c) SS | P(EO)$_{50}$Cu(CF$_3$SO$_3$)$_2$ |SS, and (d) Cu | P(EO)$_{50}$Cu(CF$_3$SO$_3$)$_2$ | Cu. The temperature is 85°C.

EuBr$_3$,[59] PEO–Nd(CF$_3$SO$_3$)$_2$,[60] MEEP–Nd$_2$(SO$_4$)$_3$, and MEEP–Gd$_2$(SO$_4$)$_3$.[22] Of these, the lanthanum systems have been the most extensively studied. The morphology of PEO–La(ClO$_4$)$_3$ appears to be simpler than that of many of the divalent materials in that only one high-melting (>200°C) crystalline complex forms. A eutectic near an O:La ratio of 10:1 and melting around 50°C were detected by differential scanning calorimetry analysis. The conductivity measurements are summarized in Figure 7.15. Low conductivities at high salt content are probably the result of large quantities of crystalline complex. Alternating-current impedance measurements on a La|P(EO)$_{12}$La(ClO$_4$)$_3$|La cell at 120°C showed typical blocking behavior. Application of a 10-mV dc potential resulted in the current decaying to zero, and it may be concluded that the La^{3+} ion is largely immobile in high-molecular-weight PEO.

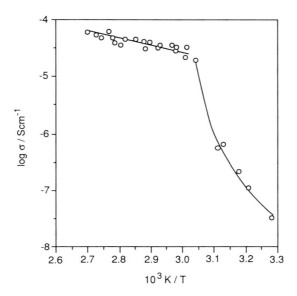

Figure 7.14. Temperature dependence of the conductivity for a $P(EO)_{50}Cu(CF_3SO_3)_2$ electrolyte.

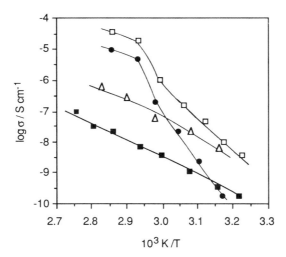

Figure 7.15. Temperature dependence of the conductivity for $P(EO)_xLa(ClO_4)_3$ electrolytes. $x = 50$ ($\square$), 20 ($\bullet$), 12 ($\triangle$), and 8 ($\blacksquare$).

References

1. F. Defendini, M. B. Armand, W. Gorecki, and C. Berthier, *Proceedings, Electrochemical Society Meeting, San Diego* (1986), p. 893.

2. P. Donoso, W. Gorecki, C. Berthier, F. Defendini, and M. B. Armand, *First International Symposium on Polymer Electrolytes (ISPE-1), St Andrews* (1987), paper 17.

3. P. Donoso, W. Gorecki, C. Berthier, F. Defendini, C. Poinsignon, and M. B. Armand, *Solid State Ionics* **28–30** (1988), 969.

4. M. Armand and M. Gauthier, in *High Conductivity Solid Ionic Conductors. Recent Trends and Applications* (T. Takahashi, Ed.), World Science, Singapore (1989), p. 114.

5. J. C. Lassegues, B. Desbat, O. Trinquet, F. Cruege, and C. Poinsignon, *Solid State Ionics* **35** (1989), 17.

6. F. Defendini, Ph.D. thesis, Grenoble (1987).

7. D. Pedone, Ph.D. thesis, Grenoble (1987).

8. M. F. Daniel, B. Desbat, and J. C. Lassegues, *Solid State Ionics* **28–30** (1988) 632.

9. R. Tanaka, T. Iwase, T. Hori, and S. Saito, *First International Symposium on Polymer Electrolytes (ISPE-1), St Andrews* (1987), paper 31.

10. M. Watanabe, R. Ikezawa, K. Sanui, and N. Ogata, *Macromolecules* **20** (1987), 968.

11. R. P. Singh, P. N. Gupta, S. L. Agrawal, and U. P. Singh, in *Solid State Ionics,* Vol. **135** (G. Nazri, R. A. Huggins, and D. F. Shriver, Eds.), Materials Research Soc., Pittsburgh (1989), p. 361.

12. A. Polak, S. Petty-Weeks, and A. J. Beuhler, *Chem. Eng. News* **28,** (1985).

13. H. Schmidt, *J. Non-Cryst. Solids* **100** (1988), 51.

14. H. Schmidt, *ACS Symp. Series (Inorg. Organomet. Polym.)* **360** (1988), 333.

15. M. Popall and H. Schmidt, *TEC,* Grenoble (1988).

16. D. Ravaine, A. Seminal, Y. Charbouillot, and M. Vincens, *J. Non-Cryst. Solids* **82** (1986), 210.

17. Y. Charbouillot, D. Ravaine, M. Armand, and C. Poinsignon, *J. Non-Cryst. Solids* **103** (1988), 325.

18. F. Rousseau, M. Popall, H. Schmidt, C. Poinsignon, and M. Armand, in *Second International Symposium on Polymer Electrolytes* (B. Scrosati, Ed.), Elsevier, London (1990), p. 325.

19. Y. Charbouillot, C. Poinsignon, D. Ravaine, and A. Domard, *First International Symposium on Polymer Electrolytes (ISPE-1), St. Andrews* (1987), paper 15.

20. Y. Charbouillot, C. Poinsignon, and M. B. Armand, *Proceedings, 6th International Conference on Solid State Ionics, Garmisch-Partenkirchen* (1987), p. 189.

21. C. Poinsignon, *Mater. Sci. Eng.* **B3** (1989), 31.

22. P. M. Blonsky, D. F. Shriver, P. Austin, and H. R. Allcock, *Solid State Ionics* **18/19** (1986), 258.

23. G. C. Farrington and R. G. Linford, in *Polymer Electrolyte Reviews—2* (J. R. MacCallum and C. A. Vincent, Eds.), Elsevier Applied Science, London (1989), p. 255.

24. A. Patrick, M. Glasse, R. Latham, and R. G. Linford, *Solid State Ionics* **18/19** (1986), 1063.

25. R. Neat, M. Glasse, R. G. Linford, and A. Hooper, *Solid State Ionics* **18/19** (1986), 1088.

26. R. Huq and G. C. Farrington, *J. Electrochem. Soc.* **135** (1988), 524.

27. P. G. Bruce, F. Krok, and C. A. Vincent, *Solid State Ionics* **27** (1988), 81.

28. R. Huq, G. Chiodelli, P. Ferloni, A. Magistris, and G. C. Farrington, *J. Electrochem. Soc.* **134** (1987), 364.

29. R. Huq and G. C. Farrington, *Solid State Ionics* **28–30** (1988), 990.

30. R. Huq, M. A. Salzberg, and G. C. Farrington, *J. Electrochem. Soc.* **136** (1989), 1250.

31. F. M. Gray, *Eur. Polym. J.* **24** (1988), 1009.

32. F. M. Gray, C. A. Vincent, P. G. Bruce, and J. Nowinski, in *Second International Symposium on Polymer Electrolytes* (B. Scrosati, Ed.), Elsevier, London (1990), p. 299.

33. P. G. Bruce, F. Krok, J. Evans, and C. A. Vincent, *Br. Polym. J.* **20** (1988), 193.

34. M. Eigen, *Pure Appl. Chem.* **6** (1963), 97.

35. L. L. Yang, A. R. McGhie, and G. C. Farrington, *J. Electrochem. Soc.* **133** (1986), 1380.

36. R. G. Pearson, *J. Chem. Educ.* **45** (1968), 581.

37. M. Cole, Ph.D. thesis, Leicester Polytechnic, Leicester (1989).

38. A. Gilmour, M. Z. A. Munshi, B. B. Owens,, and W. H. Smyrl, in *Materials and Processes for Lithium Batteries* (K. M. Abraham and B. B. Owens, Eds.), Electrochem. Soc., 89-4 (1989), p. 358.

39. L. L. Yang, R. Huq, and G. C. Farrington, *Solid State Ionics* **18/19** (1986), 291.

40. K. C. Andrews, M. Cole, R. J. Latham, R. G. Linford, H. M. Williams, and B. R. Dobson, *Solid State Ionics* **28–30** (1988), 929.

41. G. C. Farrington, H. Yang, and R. Huq, in *Solid State Ionics,* Vol. **135** (G. Nazri, R. A. Huggins, and D. F. Shriver, Eds.), Materials Research Soc., Pittsburgh (1989), p. 319.

42. R. Huq and G. C. Farrington, in *Solid State Ionic Devices* (B. V. R. Chowdari and S. Radhakrishna, Eds.), World Science, Singapore (1988), p. 87.

43. W. C. Bermudez, T. M. Abrantes, J. Morgado, and L. Alcacer, in *Second International Symposium on Polymer Electrolytes* (B. Scrosati, Ed.), Elsevier, London (1990), p. 251.

44. T. M. A. Abrantes, L. J. Alcacer, and C. A. C. Sequeira, *Solid State Ionics* **18/19** (1986), 315.

45. M. Cole, R. J. Latham, R. G. Linford, W. S. Schlindwein, and M. H. Sheldon, in *Solid State Ionics,* Vol. **135** (G. Nazri, R. A. Huggins, and D. F. Shriver, Eds.), Materials Research Soc., Pittsburgh (1989), p. 383.

46. A. K. Saraswat, A. Magistris, G. Chiodelli, and P. Ferloni, *Electrochim. Acta* **34** (1989), 1745.

47. P. G. Bruce, F. Krok, J. Nowinski, F. M. Gray, and C. A. Vincent, *Mater. Res. Forum* **42** (1989), 193.

48. B. Scrosati, in *Solid State Ionic Devices* (B. V. R. Chowdari and S. Radhakrishna, Eds.), World Science, Singapore (1988), p. 133.

49. F. Bonino, S. Pantaloni, S. Passerini, and B. Scrosati, *J. Electrochem. Soc.* **135** (1988), 1961.

50. S. Passerini, R. Curini, and B. Scrosati, *Appl. Phys. A* **49** (1989), 425.

51. K. Singh, G. Chiodelli, A. Magistris, and P. Ferloni, in *Second International Symposium on Polymer Electrolytes* (B. Scrosati, Ed.), Elsevier, London (1990), p. 291.

52. A. Magistris, G. Chiodelli, K. Singh, and P. Ferloni, *Solid State Ionics* **40/41** (1990), in press.

53. G. K. Jones, G. C. Farrington, and A. R. McGhie, in *Second International Symposium on Polymer Electrolytes* (B. Scrosati, Ed.), Elsevier, London (1990), p. 239.

54. B. Scrosati, personal communication.

55. F. M. Gray and C. A. Vincent, unpublished results.

56. F. M. Gray, C. A. Vincent, and B. Scrosati, to be published.

57. J. E. Weston and B. C. H. Steele, *Solid State Ionics* **2** (1981), 347.

58. P. G. Bruce, F. M. Gray, and C. A. Vincent, *Solid State Ionics* **38** (1990), 231.

59. R. Huq and G. C. Farrington, in *Second International Symposium on Polymer Electrolytes* (B. Scrosati, Ed.), Elsevier, London (1990), p. 273.

60. A. S. Reis Machado and L. Alcacer, in *Second International Symposium on Polymer Electrolytes* (B. Scrosati, Ed.), Elsevier, London (1990), p. 283.

CHAPTER 8

Transport Properties: Effects of Dynamic Disorder

Polymer electrolyte materials are characterized by the presence of static disorder arising from the disordered polymer host. Above the glass transition temperature, T_g, there is in addition dynamic disorder; that is, the local environment at any one point in the material undergoes substantial change with time as a result of the liquidlike motion that takes place. This type of disorder has an important role in determining charge transport in these materials. It is also important to realize that ionic motion takes place above T_g where the material is *locally* fluid. The overall solid-state properties are due largely to chain entanglements and to relatively slow relaxation times, which result in minimal long-range flow. Thus, the systems that are dealt with here can be considered as extremely viscous fluids where fluidlike behavior should dominate the transport process, although an ion-hopping mechanism, characteristic of solids, may also contribute to ion transport in polymer electrolytes.

8.1. Macroscopic Models

To understand the temperature and pressure dependence of the dc conductivity in polymer electrolytes, the conductivity, σ, can be expressed in general terms as

$$\sigma(T) = nq\mu$$

where q is the charge, μ is the mobility, and n is the number of carriers. By making the assumption that the number of charge carriers is invariant and that the charge is unity, the conductivity can be related directly to the mobility of the charged species. It is thus possible to develop models that describe the conductivity solely in terms of solvent fluidity and this is a starting point in the understanding of ion transport in real polymer electrolyte systems.

It is known that for transport in liquidlike systems, the temperature dependence of a number of relaxation and transport processes in the vicinity of the glass transition temperature can be described by the Williams–Landel–Ferry (WLF) equation.[1] This relationship was originally derived solely by fitting observed data for a number of liquid systems. It expresses a characteristic property, for example, fluidity, reciprocal dielectric relaxation time, or magnetic resonance relaxation rate, in terms of a shift factor, a_T, which is the ratio of any mechanical relaxation process at temperature T to its value at a reference temperature T_s and is defined by

$$\log a_T + \text{const.} = \log \left[\frac{\eta(T)}{\eta(T_s)} \right]$$

$$= -\frac{C_1(T-T_s)}{(C_2+T-T_s)} \tag{8.1}$$

where η is the viscosity, T_s is a reference temperature, and C_1 and C_2 are constants that may be obtained experimentally. The constants have so-called "universal" values, are independent of the measured property, and take the values $C_1 = 8.9$ K and $C_2 = 102$ K. T_0, although arbitrary, is often taken to be 50 K above the glass transition temperature. The shift factor may therefore be written as

$$\log a_T = \frac{-17.4 \, (T-T_g)}{(51.6+T-T_g)} \tag{8.2}$$

Because T_g measurements are kinetically determined, however, this is a less accurate form of the equation. Shift factors can be measured directly and can be defined for different properties, for example, viscosity. Very often it is observed that the shift factors are independent of the measured property and also, if for every polymer system a different reference temperature T_s is chosen, and a_T is expressed as a function of $(T - T_s)$, then a_T turns out to be nearly universal for all polymers. Williams, Landel, and Ferry[1] believed that the universality of the shift factor was due to a dependence of relaxation rates on free volume. Although the relationship has *no free-volume basis*, the constants C_1 and C_2 may be given significance in terms of free-volume theory.[2] The temperature dependence of the conductivity may be fitted to a WLF-type relationship in that decrease in the T_g leads to an increase in conductivity. Measurements of shift factors have been carried out by Cheradame and co-workers[3–9] on crosslinked polymer electrolyte networks by measuring the mechanical loss tangent. Figure 8.1 shows values of $\log a_T$ for PEO-based crosslinked networks as a function of $T - T_g$. The fitted line corresponds to $C_1 = 10.5$ and $C_2 = 100$. The shift factors were shown to correlate with ionic conductivity (Figure 8.2), suggesting that ionic motion was promoted by polymer segmental motion.

   Currently most theories for polymer dynamics are based on one of two assumptions. The first put forward by Cohen and Turnbull[10] attempts to explain behavior in liquid systems by assuming a model of a continuously changing free-volume distribution. In practice, the application of free-volume ideas to conductivity in polymer electrolytes has been carried out largely through application of the Vogel–Tammann–Fulcher (VTF)[11] form of the equation (of which the WLF relation is an extension). This is again an empirical equation, and in its original form,

$$\eta^{-1}(T) = A \, \exp^{-B/(T-T_0)} \tag{8.3}$$

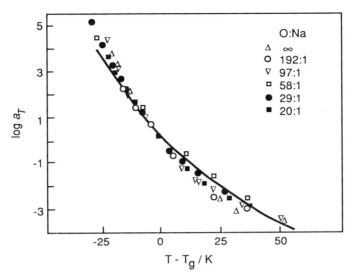

Figure 8.1. Shift factor, a_T, for PEO(400)-based network electrolytes as a function of reduced temperature, $T-T_g$. Reprinted with permission from Ref. 7, *IUPAC Macromolecules,* Copyright 1982, Pergamon Press Plc.

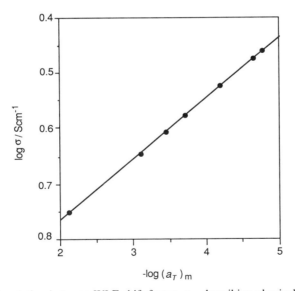

Figure 8.2. Correlation between WLF shift factor, a_T, describing physical relaxation, and the conductivity for PEO networks. The direct proportionality with slope of unity implies conductivity to be inversely proportional to viscosity as assumed by the Stokes–Einstein relationship or Walden relationship for fully dissociated electrolytes. Reproduced from Ref. 8 with permission of the publisher, Hüthig & Wepf Verlag, Basel.

Here T_0 is $T_s - C_2$ and A is a prefactor determined by the transport coefficient (in this case η^{-1}) at the given reference temperature and A is proportional to $T^{1/2}$. The constant B is an energy divided by the Boltzmann constant and is not related to any simple activation process.[2] The above equation holds for many transport properties and, by assuming the electrolyte to be fully dissociated in the solvent, it can be related to the diffusion coefficient through the Stokes–Einstein equation giving the form to which the conductivity, σ, in polymer electrolytes is often fitted:

$$\sigma = \sigma_0 \exp^{-B(T-T_0)} \tag{8.4}$$

The form of this equation suggests that thermal motion above T_0 contributes to relaxation and transport processes, and that for low T_g, faster motion and faster relaxation should be observed.

Equation (8.4) may be expressed in a number of slightly different forms that depend on the model and assumptions made in the original derivation. If ionic diffusion is considered to be an activated process as, for example, in the case of glasses and ceramics, then included in the preexponential term of Eq. (8.4) is the attempt frequency, ν_0, for ion mobility. Several expressions have been proposed for the attempt frequency but in the simplest case it may be assumed to be temperature independent and ν_0 is a constant. In transition-state theory, it is presumed that ν_0 represents a fully excited vibrational degree of freedom so that $\nu_0 = kT/h$. Following the approach of Cohen and Turnbull,[10] one might imagine that the ion moved about its cage of neighboring atoms with the velocity of an ideal gas molecule. This kinetic theory gives $\nu_0 = (8kT/\pi m)^{1/2}/2d$. Thus the preexponential factor of Eq. (8.4) may be written as σ_0/T^m, where $m = 0, 1,$ or $\frac{1}{2}$. Interestingly, for a number of glassy materials, the conductivity variation over a large temperature range (310°C) has been found to best fit the equation where the preexponential factor is temperature independent.[12] In practice, these models are not applicable to ion transport in polymer electrolytes and, indeed, it should be emphasized that not all systems follow the type of behavior described by Eq. (8.4).[13–15] In addition, the temperature range over which measurements are made is never large enough to ascertain whether the inclusion of a temperature term in the preexponential factor is appropriate or not. There is no advantage therefore in using a refined form of the equation when the simpler form of Eq. (8.4) gives a satisfactory fit to the data.

In the free-volume model, motion is assumed to be a non-thermally activated process but occurs as a result of redistribution of the free volume. A distribution function for the size of voids in the given material is derived, and the probability of this distribution is maximized. With viscosity proportional to the inverse of the diffusion coefficient from the Stokes–Einstein equation (assuming a strong electrolyte) and nearly independent of temperature at constant volume,[16] and if the transport or relaxation rate is determined by the rate at which voids of certain minimum critical volume v^* are created, then

$$\eta^{-1} \sim \exp^{-\gamma v^*/v_f} \tag{8.5}$$

where v_f is the free volume per mole, v^* is the critical volume per mole, and γ is a constant that allows for overlap of the free volume. If v_f is expanded around a temperature T_0, the temperature at which the free volume vanishes is,

$$v_f = v_f(T_0) + (T-T_0) \left(\frac{\partial v_f}{\partial T_0} \right)_{T_0} \tag{8.6}$$

It may be shown[2] that by assumption of the Nernst–Einstein relationship

$$\sigma = \frac{nq^2 D}{kT} \tag{8.7}$$

with n and q the carrier concentration and charge, respectively, and D the diffusion coefficient, a free-volume expression for the conductivity in the form of Eq. (8.4) may be derived, provided the electrolyte is fully dissociated. Also,

$$B = \gamma v^* / \left(\frac{\partial v_f}{\partial T} \right)_{T_0} \tag{8.8}$$

The constant B is related to the critical free volume for transport v^* and to expansivity; it is not a simple activation barrier. In polymer electrolytes, v^* is generally taken as fixed by the size of the polymer segment rather than the motion of the ion, as the polymer strands must move before either cations or anions can be transported. Miyamoto and Shibayama[17] proposed a model that may be seen as an extension to free-volume theory and allows explicitly for the energy requirements of ion motion relative to counterions and polymer host. This was elaborated by Cheradame and co-workers[3–9] to describe ionic conductivity in crosslinked networks. The conductivity was expressed in the form

$$\sigma = \sigma_0 \exp\left[-\frac{\gamma v^*}{v_f} - \frac{\Delta E}{RT} \right] \tag{8.9}$$

which is a combination of Arrhenius and free-volume behavior. The activation energy, ΔE, is given as

$$\Delta E = E_j + W/2\epsilon \tag{8.10}$$

where E_j is the energy barrier for cooperative ion transfer from one hole site to another, W is the dissociation energy of the salt, and ϵ is the dielectric constant of the matrix. If ΔE is relatively small, then the second term in Eq. (8.9) tends to unity and free-volume behavior predominates. By reducing the data so as to eliminate the free-volume behavior, the first term in Eq. (8.9), Cheradame[7] deduced an activation energy $\Delta E \sim 20$ kJ mol^{-1} for sodium tetraphenyl borate-doped PEO networks. Cheradame showed that the reduced conductivity is a linear function of the WLF shift factor:

$$\ln \left[\frac{\sigma(T)}{\sigma(T_g)} \right] = - \left(\frac{v_i^*}{v_p^*} \right) \ln a_T \tag{8.11}$$

where v_i^* and v_p^* are the critical free volumes for ions and polymer segment motion, respectively. v_i^*/v_p^* was not found to vary greatly from unity, suggesting that both properties relate to the same basic process. Watanabe and co-workers[18–23] have also studied the conductivity and elastic modulus of salt-containing polyether networks and have related these in a similar manner; however, when the conductivities were plotted as a function of salt concentration at various reduced temperatures, the increase in conductivity was larger than that predicted on the assumption of complete dissociation. They considered that this was a consequence of the fact that ΔE in Eq. (8.9) was not negligible. It was shown that the critical volumes for conductivity and mechanical relaxation were virtually identical and that ΔE could be extracted from

$$\frac{T}{T_g}\left[\frac{\sigma(T)}{\sigma(T_g)} \Big/ \frac{1}{a_T}\right] = A \exp\left(\frac{-\Delta E}{RT}\right) \tag{8.12}$$

The free-volume model and extensions to deal with such processes as ion association have been very extensively used in polymer electrolyte studies. Although successful in rationalizing many of the properties of these materials, in particular the temperature dependence of conductivity, they are inadequate for several reasons. The major weakness is that the model ignores the kinetic effects associated with macromolecules. In addition, the model does not relate directly to a microscopic picture and therefore does not predict straightforwardly how such variables as ion size, polarizability, ion pairing, solvation strength, ion concentration, polymer structure, or chain length will affect the conduction process.

The second group of theories is based on the configurational entropy model of Gibbs and co-workers[24,25] which goes some way to overcoming these deficiencies. As for free-volume models, the configurational entropy model discusses only the properties of the polymer. The mass transport mechanism in this model is assumed to be a group cooperative rearrangement of the chain, giving the average probability of a rearrangement as

$$W = A \exp\left(\frac{-\Delta\mu S_c^*}{kTS_c}\right) \tag{8.13}$$

where S_c^* is the minimum configurational entropy required for rearrangement and is often taken as $S_c^* = k \ln 2$,[2] S_c is the configurational entropy at temperature T, and $\Delta\mu$ is the free energy barrier per mole that opposes the rearrangement. As the glass transition temperature is approached, the molecular relaxation time becomes increasingly longer and equilibrium cannot be maintained. Eventually the dynamic configurational entropy will reach zero at a temperature T_0, often found to be approximately 50 K below T_g. If W is related to the inverse relaxation time and ΔC_p, the heat capacity difference between the liquid and glass state, is assumed to be temperature independent, a WLF-type relationship can be given:

$$-\log a_T = \frac{a_1(T-T_s)}{(a_2+T-T_s)} \tag{8.14}$$

The constants a_1 and a_2 correspond to

$$a_1 = 2.303\Delta\mu S_c^* / \Delta C_p kT_s \ln\left(\frac{T_s}{T_0}\right)$$

$$a_2 = T_s \ln\left(\frac{T_s}{T_0}\right) \Big/ \left[1 + \ln\left(\frac{T_s}{T_0}\right)\right]$$

with a_2 slightly temperature dependent. A VTF form can also be written:

$$W = A \exp[-K_\sigma/(T - T_0)] \tag{8.15}$$

with $K_\sigma = \Delta\mu S_c^*/k\Delta C_p$. Shriver and co-workers[26,27] applied the configurational entropy model to polymer electrolytes. They used the assumption[28,29] that $\Delta C_p = B/T$, where B is a constant, and showed that

$$\sigma(T) = A \exp[-K_\sigma/(T - T_0)] \tag{8.16}$$

with $K_\sigma = \Delta\mu S_c^* T_0/kB$. The approach appears to have merit, and analyses of several polymer electrolyte systems[26,30–34] give reasonable values for the activation energy and values close to 50 K for $T_g - T_0$. Certain aspects of conductivity are also implied by the configurational entropy approach, for example, the fall in conductivity with increasing pressure, the molecular weight independence of conductivity in amorphous polymers above a certain molecular weight, and the fact that if T_0 is lowered, the conductivity at a fixed reduced temperature $(T - T_0)$ should increase. Angell and Bressel[35] have rationalized the conductivity maxima in conductivity/concentration plots on the basis of an isothermal version of the VTF equation

$$\sigma = AX \exp[-K/Q(X_0 - X)] \tag{8.17}$$

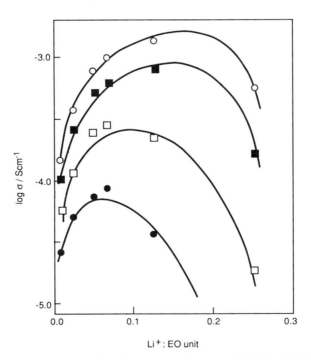

Figure 8.3. Variation of log σ with salt concentrations for PVMEO$_3$–LiClO$_4$ systems at 50°C (●), 75°C (□), 100°C (■), and 120°C (○). The solid lines were calculated from Eq. (8.17). From Ref. 30. Reproduced by permission of the SCI.

where $Q = dT_g/dX$, X is the mole fraction of salt, and X_0 is the mole fraction of salt at $T = T_g$. Cowie et al.[30] found reasonably good agreement between experimental data for an amorphous comb-type polymer–LiClO$_4$ electrolyte and the conductivity and maxima predicted by Eq. (8.17) (Figure 8.3). The preexponential factor, A, was found to decrease with increasing temperature to a particular temperature and then show a slight fall. This may imply some temperature dependence of the cation-to-polymer interaction equilibrium.

It is almost always possible to obtain a reasonable fit of experimental data for amorphous materials to a VTF/WLF-type equation; however, as pointed out by Greenbaum et al.,[36] significant discrepancies can arise in parameters if the fittings are made over different temperature ranges. It has also been noted[34,36,37] that, in general, Arrhenius-type temperature dependence of the conductivity, rather than VTF-type behavior, is observed well above the T_g. Similar behavior, however, is also noted close to T_g.[34,38]

8.2. Microscopic Approach

Although VTF and WLF equations along with free-volume or configurational entropy approaches offer many advantages, that is, they describe adequately many of the transport properties in polymer electrolytes, they are not based on a microscopic treatment and therefore local mechanistic information is lost.

Angell and Torell[39–41] proposed that a comparison of structural relaxation time, τ_s, and conductivity relaxation time, τ_σ, could characterize transport mechanisms in solid ionic conductors. A decoupling index, R_τ, was defined where

$$R_\tau = \tau_s/\tau_\sigma \qquad (8.18)$$

The behavior of R_τ as a function of temperature has been investigated in different ionic materials. In glassy materials such as LiAlSiO$_4$, R_τ is very near unity at high temperatures well above the glass transition temperature. As the molten glass is cooled below T_g, R_τ falls out of equilibrium at a particular temperature and the structural relaxation time becomes progressively longer. At temperatures well below T_g, the ratio R$_\tau$ in these glasses can be very large, typically 10^{12}. This is in sharp contrast to polymer electrolytes for which the operational temperatures are above T_g. In these materials, transport and relaxation are very closely related to one another, with R_τ close to unity, suggesting very similar mechanisms for conductivity and structural relaxation. For concentrated polymer–salt complexes, R_τ is generally found to be less than unity, often of the order of 0.1. This suggests that relaxation processes that do permit rearrangement of the polymeric structure do not necessarily permit ions to move. This could arise from interionic interactions, resulting in ion immobilization or Coulomb drag. Figure 8.4 shows R_τ data for a number of compounds and demonstrates the disparity between solvent-assisted motion in the polymer electrolytes and activated hopping in the vitreous material.

Motion of ions in polymer electrolytes is strongly dependent on segmental motion of the polymer host. Based on this and assuming a weak dependence of the conductivity on interionic interactions, Druger and co-workers proposed a microscopic

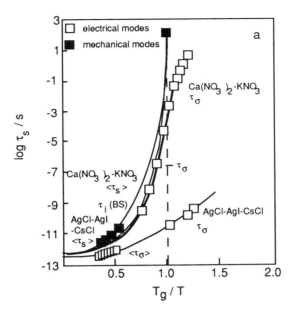

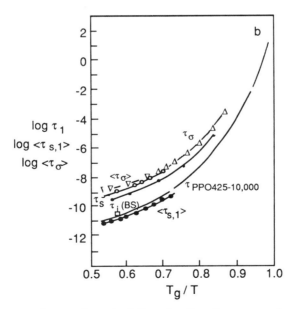

Figure 8.4. Average shear and conductivity relaxation times, τ_s and τ_σ, for ionically conductive materials. (**a**) For glassy materials $\tau_\sigma \sim \tau_s$ above T_g and $\tau_\sigma \ll \tau_s$ below T_g. (**b**) For polymer electrolytes, $\tau_\sigma \geq \tau_s$ for the relevant range $T > T_g$. Data in (**b**) are for $P(PO)_{13}NaCF_3SO_3$, and the broken line is for pure PPO. From Ref. 41. Reproduced by permission of the SCI.

model[2,42-47] to describe the transport mechanism. This is known as the dynamic bond percolation (DBP) theory. For conductivity in polymer electrolytes, cation motion and anion motion are considered to be fundamentally different. The former is visualized as the making and breaking of coordinate bonds with motion between coordinating sites, whereas anion motion is regarded as a hopping between an occupied site and a void tha' is large enough to contain the ion. Conductivity is visualized as being the result of a combination of cooperative motion with the occasional independent ion movement, the time scale for the latter much shorter than for polymer relaxation.

The simplest model involves ion hopping between sites on a lattice (not fixed as in a solid electrolyte such as AgI), with the ions obeying a hopping-type equation

$$\dot{P}_i = \sum_j P_j W_{ji} - P_i W_{ij} \tag{8.19}$$

P_i is the probability of finding a mobile ion at site i, and W_{ij} is the probability per unit time that an ion will hop from site j to site i and is equal to zero except between neighboring sites. It is then assumed that W_{ij} may take two values. One is w, the rate of ion hopping from a filled to a vacant site or zero, if all sites are already filled. Jumps are available with a relative probability f ($0 \leq f \leq 1$). Because of polymer motion, the configuration is continually changing and sites move with respect to each other. Hopping probabilities therefore readjust or renew their values on a time scale τ_{ren}, which is determined by the polymer motion. The W_{ij} values are thus fixed by the parameters w, f, and τ_{ren}. These can be related to system parameters such as ion size, free volume, temperature, and pressure. In a comparison with free-volume theory, Druger et al.[42] identified f with $\exp(-\gamma v^*/v_f)$ and w with the ion velocity divided by a lattice constant. τ_{ren} is the characteristic relaxation time corresponding to configurational or orientational changes. A number of results follow from this model:

1. For an observation time much longer than τ_{ren}, with $f > 0$, ionic motion is always diffusive.[43,44,47]
2. For very fast renewal times, motion corresponds to hopping in a homogeneous system with an effective hopping rate wf. If the renewal rate is very slow, then observed motion is that of a static bond percolation model. As the renewal time increases, diffusion coefficients and therefore conductivity increase until the rate-determining step changes to that of ion hopping and ion transport is independent of segmental motion.
3. Frequency-dependent properties such as spectroscopic behavior, dielectric relaxation, and frequency-dependent conductivity may be described by a DBP model. In Figure 8.5, substantially different behavior for the mean polymer host and the polymer ion conductor is observed. At frequencies above approximately 10 GHz, these two responses are essentially identical and may reflect displacive motion of ions or dipoles within the polymer host. For lower frequencies, only the ions continue to respond diffusively over lengthening time scales. Thus the response of the pure polymer falls to zero, whereas that of the electrolyte levels

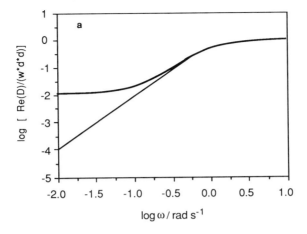

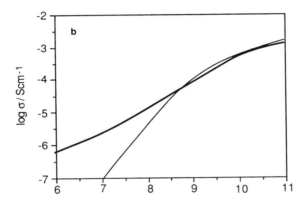

Figure 8.5. (a) Calculated frequency-dependent conductivity for a simple dynamic percolation model. Lower line represents the diffusion coefficient without renewal; bold line represents that with renewal. (b) Frequency-dependent conductivity for pure PEO (bold) and PEO–NaSCN at 22°C. Only ions are able to diffuse a long distance, corresponding to renewal diffusion. From Ref. 48.

out at low frequency, corresponding to the dc diffusion arising from the renewal process. These results are compared with experimental findings.[48]

In addition to specific applications, the DBP model has been extended to focus on the importance of lattice considerations. The role of correlations among different renewal processes and the effects of renewal on diffusion at the percolation threshold transition have been highlighted.[44–47,49–52] These have been discussed recently by Ratner and Nitzan.[53]

A number of important issues involving the structure and dynamics of ionically conducting polymers have yet to receive thorough theoretical consideration. For example, what is the nature of the cation–polymer interaction, how are the cations solvated, and what is the importance of anion–polymer interactions? More important, how do interactions between mobile ions affect diffusion and conductivity? As the ionic concentration in polymer electrolytes is usually greater than 1 mol dm^{-3} and the mean distance between ions is of the order of 0.5 to 0.6 nm, then relatively strong Coulombic interactions exist that must affect ion motion. Likewise, it questions the independent particle description of both free-volume and DBP theory. Ratner and Nitzan have recently addressed this problem.[53] They developed a model based on the transient crosslink phenomenon of polymer chains that occurs through the ionic species. This effect suppresses the configurational entropy available to the chains and is proportional to the transient crosslink density. The formation of contact ion pairs is given by the simple relationship

$$pc + a \rightleftharpoons p + ac \tag{8.20}$$

where p, c, and a represent polymer chain segment, cation, and anion, respectively. One can define, therefore, n_c^b, n_c^f, and n_p^b, as the number of coordinated (bound) cations, number of free cations, and number of coordinated polymer chain segments, with the total number of polymer chain segments, N_p, and total number of cations, N_c (equal to the number of anions), such that

$$N_p = n_p^f + n_p^b$$
$$N_c = n_c^b + n_c^f \tag{8.21}$$

If a free polymer segment becomes complexed, the entropy cost will be proportional to $n_p^f \ln \lambda$, where λ is the contribution to the entropy from each free polymer segment and depends on the number of configurations available to the solvent. By setting the energy to zero in the fully uncomplexed material, the energetic contribution to the free energy can be defined as

$$\Delta U_p = \epsilon n_b \tag{8.22}$$

where n_b is the number of bound cation–polymer pairs ($= n_c^b + n_p^b$) and ϵ is the energy difference on solvation of an ion pair which is expected to be negative. (For ϵ to be positive, there would be no thermodynamic driving force for a complex to form.) An expression for n_b may be found which gives the desired relationship between temperature and number of ion pairs

$$\frac{\epsilon \lambda n_b N_s \exp(\epsilon/kT)}{kT^2} = \left(\frac{\partial n_b}{\partial T}\right)\left[N_p + N_c - 2n_b + \lambda N_s \exp(\epsilon/kT)\right] \tag{8.23}$$

where N_s is the total number of sites. As ϵ is negative and $N_p + N_c - 2n_b + \lambda N_s \exp(\epsilon/kT)$ must be positive, $\partial n_b/\partial T < 0$. Thus the number of cations bound to the polymer decreases with temperature. This is in agreement with the experimental findings of Torell and Schantz[54] and Greenbaum and co-workers,[55,56] the latter showing that at high enough temperature phase separation occurs. In addition, the term involving configurational entropy reduction as a result of complexation suggests that at smaller chain lengths, ion pair formation and salting out should be reduced. No experimental data are currently available that support this prediction.

As yet, it has not been possible to amalgamate experimental data from techniques that relate to microscopic properties such as dielectric relaxation, Brillouin scattering, inelastic neutron scattering, or NMR relaxation times into dynamic disorder models. Properties such as τ_{ren} are clearly closely related to these relaxation properties but a more detailed understanding of the nature of the ionic conduction process is required before assignment of a relaxation time to a renewal process can be ventured.

8.3. Experimental Techniques Relating to Microscopic Dynamic Properties

8.3.1. Brillouin Scattering

Light scattering techniques can be used to study dynamic processes and to follow structural behavior over a wide time scale. Very short time processes ($\sim 10^{-14}$ s) observed by Raman spectroscopy give information about molecular and intramolecular motions such as bond vibration, rotation, or libration and the effects of interactions. This is discussed in Chapter 9. Brillouin scattering, on the other hand, gives information on longer-time behavior ($\sim 10^{-11}$ s) and allows dynamic processes such as structural relaxation, molecular reorientations, and ion diffusion to be probed. Acoustic properties of the scattering medium are probed through this technique at high frequencies (~ 10 GHz). In dielectric relaxation, the dynamic and relaxation behavior of a grouping of electric dipoles is measured. Brillouin scattering occurs as a result of density fluctuations that modulate the medium's dielectric constant, and the two techniques need not manifest the same mechanisms. The former process results from dipole reorientation associated with main-chain bond rotations and monitors motions associated with the glass–rubber relaxation. Brillouin scattering follows different activation parameters and evidences a more rapid process than dielectric relaxation. It is speculated that Brillouin scattering monitors a secondary or subglass relaxation due possibly to damped torsional oscillations.[57] For an isotropic medium the frequency shift of the Brillouin lines, f_B, is related to the velocity of sound, v, through

$$v = f_B \lambda_0 \,/\, 2n \sin(\theta \,/\, 2) \tag{8.24}$$

where λ_0 is the wavelength of incident light, n is the refractive index, and θ is the scattering angle. In a frequency or temperature region where relaxation events occur, the acoustic wave will be attenuated. This will lead to velocity dispersion and absorption which for a single-relaxation-time process are

$$\frac{(v^2 - v_0^2)}{(v_\infty^2 - v_0^2)} = \frac{\omega^2 \tau^2}{(1 + \omega^2 \tau^2)}$$

$$\alpha = \frac{(v_\infty^2 - v_0^2)}{2v^3} \cdot \frac{\omega_2 \tau}{(1 + \omega^2 \tau^2)} \tag{8.25}$$

where v_0 and v_∞ are the low- and high-frequency limits of the velocity and α is the absorption coefficient related to the broadening of the Brillouin linewidth, Γ_B, by

$$\Gamma_B = \alpha v/2\pi \qquad (8.26)$$

In polymer electrolytes, this technique has been used to investigate the elastic properties of polymer–salt complexes and the structural response times which are related to local mobilities within the chains. It is thus possible to probe microscopic viscosity changes and relaxation times. Torell and Angell have estimated that the size of the locally relaxing unit detected by Brillouin scattering is greater than 1 but probably less than 7.[41] The technique and its application to polymer electrolytes have recently been reviewed by Torell and Schantz.[54] Brillouin scattering experiments have been carried out on a range of liquid PPO polymer electrolytes: PPO–M SCN (M = Li, Na, K),[58] PPO–NaCF$_3$SO$_3$,[59] and PPO–LiClO$_4$.[60] Although the bulk viscosity of the polymer is highly molecular weight dependent, it was found that the sound velocity remains unchanged and therefore the microscopic viscosity is independent of molecular weight. Relevant information relating to solid polymer electrolytes can thus be obtained by studying liquid analogs.

Adding a salt to a coordinating polymer not only increases the macroscopic viscosity[61] but also changes the microscopic viscosity. This apparent decrease in chain flexibility caused, it is assumed, by transient chain crosslinking by ions can be detected as a shift of the entire velocity–time plot to higher temperatures as the salt concentration is increased and, also, as a shift in the maximum of the sound absorption–temperature plot. Surprisingly, there appears to be no dependence on the cation as studies of PPO–MSCN systems (M = Li, Na, K) have shown (Figure 8.6),[58] despite the dc conductivity showing a cation dependence at fixed salt concentration and temperature.[62–64] This seems to imply that the variable responsible for the cation effect is the number of carriers per unit volume. For M = Li, Na, and K, the lattice energies for MSCN are 806, 682, and 616 kJ mol^{-1}, respectively, and

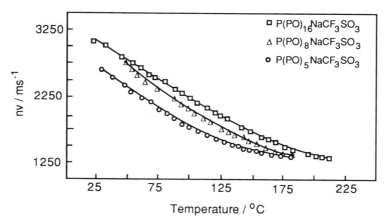

Figure 8.6. Temperature dependence of the hypersonic velocity for PPO–NaCF$_3$SO$_3$ systems. From Ref. 54.

therefore the cation–anion attractive forces are likely to be least pronounced for KSCN systems and account for the higher conductivity values observed.

Information on structural relaxation processes at Brillouin frequencies can also be obtained by analysis of the relaxing part of the longitudinal modulus, M^*. Real and imaginary parts of the elastic modulus are related to experimental data for sound velocity and absorption through the following relations, written in the reduced N' and N'' forms,

$$N'(\omega) = \frac{M'(\omega) - M_0(\omega)}{M_\infty(\omega) - M_0(\omega)} = \frac{v^2(\omega) - v_0^2(\omega)}{v_\infty^2(\omega) - v_0^2(\omega)}$$

$$N''(\omega) = \frac{M''(\omega) - M_0(\omega)}{M_\infty(\omega) - M_0(\omega)} = \frac{2\alpha(\omega)v^3(\omega)}{\omega[v_\infty^2(\omega) - v_0^2(\omega)]} \tag{8.27}$$

and can be compared with the analytical expressions

$$N'(\omega') = \omega\tau \int d(t/\tau)\sin(\omega\tau)\phi(t/\tau)$$

$$N''(\omega) = \omega\tau \int d(t/\tau)\cos(\omega\tau)\phi(t/\tau) \tag{8.28}$$

where τ is the relaxation time. The relaxation time function $\phi(t)$ has the form of a Williams–Watts expression[65]

$$\phi(t) = \exp[-(t/\tau)^\beta] \tag{8.29}$$

where $0 < \beta \le 1$ and is a measure of the distribution of relaxation times, equal to one for a single relaxation time. To compare experimental and analytical data and to determine relaxation parameters τ and β, it has been assumed that the temperature dependence of the relaxation time follows an equation of the VTF form[59,66]:

$$\ln \tau = \ln \tau_0 + B/(T - T_0) \tag{8.30}$$

Results for PPO systems are shown in Figure 8.7. The modulus plots for the various PPO–NaCF$_3$SO$_3$ electrolytes are superimposed at a single reduced temperature $(T - T_0)$. The best fit between experimental and analytical data for PPO gives $\beta = 0.4$, $\tau_0 = 3.46 \times 10^{-14}$ s, $B = 1169$ K, and $T_0 = 170$ K.[66] NaCF$_3$SO$_3$-complexed PPO shows a similar but slightly broader distribution with $\beta = 0.35$.[58] Values of T_0 were assumed to be $T_g - 30$ K, as found for pure PPO[67]; that is, $T_0 = 205$, 239, and 252 K for O:Na ratios of 16:1, 8:1, and 5:1, respectively, with the corresponding B parameters 1098, 1114, and 1208 K. Using the VTF function, various relaxation time–temperature relations can be generated as shown in Figure 8.8. The combined data from photon correlation,[68] Brillouin scattering,[66] and Raman studies[54,60] on PPO infer a rather remarkable applicability of Eq. (8.30) in describing the relaxation behavior over a very broad time scale, from the T_g through to the high-temperature limit $(1/T = 0)$. The limiting value for τ of 2×10^{-14} s was evaluated from Raman data[60] and was found to be independent of salt concentration. It is also very close to the value of τ_0 given earlier. The effect of increasing salt

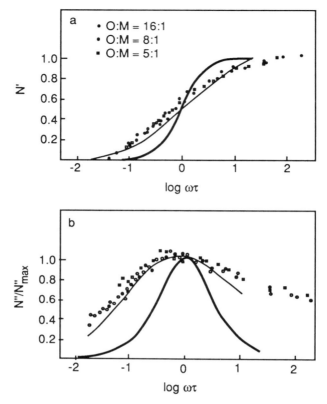

Figure 8.7. (a) Real N' and (b) imaginary N'' parts of the normalized longitudinal modulus of PPO–NaCF$_3$SO$_3$ systems. For comparison, results for pure PPO (solid lines) are shown. Bold lines represent single-relaxation-time behavior. From Ref. 54.

concentration is to move the relaxation curves to longer times but to retain this upper temperature limit.

8.3.2. Dielectric Relaxation

Frequency dependence studies of the conductivity and of the dielectric response may be informative in distinguishing detailed mechanisms for ion transport and differences in ion–polymer interactions in polymer electrolytes. Dielectric relaxation gives a measure of the dynamic and relaxation behavior of the electric dipoles in the matrix. For PEO, relaxations are molecular weight dependent, presumably because of different degrees of crystallinity.[69] High-molecular-weight PEO shows mechanically three relaxations[70] designated the α, β, and γ transitions because of their appearance at high temperature when relaxation is studied as a function of temperature at constant frequency. In intermediate- to low-molecular-weight mate-

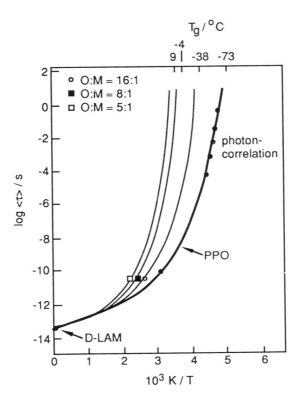

Figure 8.8. Log $\langle \tau \rangle$ versus inverse temperature for PPO–NaCF$_3$SO$_3$ systems. The results for pure PPO are also shown for comparison. From Ref. 54.

rial, the α and β processes merge. Dielectrically, however, only the β and γ processes appear to be active,[69,70] although Fontanella and co-workers have observed the α relaxation by the thermally stimulated depolarization (TSDC) method.[71,72] Figure 8.9 shows the temperature dependence of the dielectric loss, ϵ'', for PEO. For a single crystal, the β relaxation is missing. It thus appears that the α process can be attributed to a crystal-disordering mechanism, whereas the β process refers to relaxation in the amorphous phase. The temperature dependence of the γ peak infers it will merge with the β peak at high temperature.[73] This relaxation has been associated with an amorphous transition involving short-range bond rotations. On nomenclature, it is worth noting that often α is designated α_c, β is termed α_a, and occasionally γ is denoted by β.[74] The α, β, and γ terms will be used here.

The dielectric relaxation behavior of PPO is shown in Figure 8.10.[57] Because of the methyl side group, the dipole of each monomer unit does not precisely bisect the C—O—C bond angle, resulting in a net cumulative moment in the direction parallel to the arms of the chain, in addition to the much larger perpendicular components of the unit dipoles. This parallel component gives rise to a relaxation at a frequency

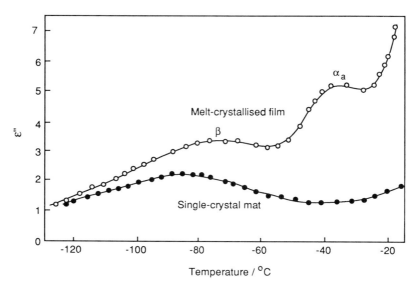

Figure 8.9. Temperature dependence of the dielectric loss factor ϵ'', for a single-crystal mat and melt-crystallized film of PEO at 12.8 kHz. From Ref. 74, Y. Ishida, *J. Polym. Sci. A2*, Copyright (1969), John Wiley & Sons, Inc. Reprinted by permission of John Wiley & Sons, Inc., All Rights Reserved.

lower (higher temperature) than that of the main peak.[75] The main dispersion peak is found to be insensitive to molecular weight and is associated with local segmental motion as in PEO.

Only a few studies of the dielectric properties of polymer electrolytes have been reported in the radio and microwave frequency regions.[76–79] These were carried out on $P(EO)_{4.5}NaSCN$, $P(EO)_{3.4}NaBH_4$,[76] $P(EO)_8NH_4CF_3SO_3$,[79] and various concentrations of PEO–RbI[77] and PEO–LiClO$_4$.[77,78] Shriver and co-workers[76,79] used a lumped circuit technique for measurements from 500 kHz to 250 MHz and employed a standing wave method at higher frequencies using a coaxial line (300 MHz–10 GHz) or a waveguide (>10 GHz) for the samples.[80] Studies by Gray et al.[77,78] used an alternative approach of time domain spectroscopy in which the system is disturbed by a fast step pulse. Transient methods are considerably less time consuming than the point-to-point approach in the frequency domain and loss of accuracy is minimal. Measurements could be performed in the range 10 MHz to 10 GHz by this technique.

Figure 8.11 shows the frequency-dependent conductivity, σ, of PEO, PEO–NaSCN, and PEO–NaBH$_4$, which may be calculated from the dielectric loss parameter (ϵ'') by

$$\sigma = \omega\epsilon_0\epsilon''$$

where ω is the angular frequency and ϵ_0 is the permittivity of free space. Above 10 GHz the materials all behave in a similar manner and the response has been at-

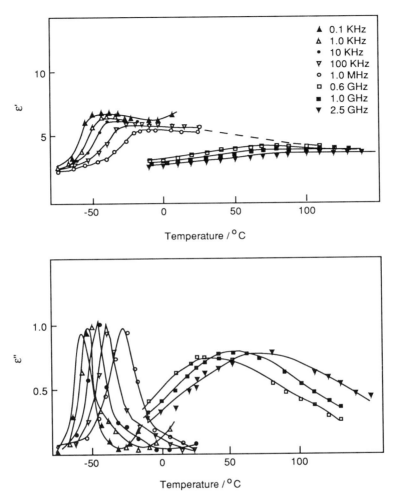

Figure 8.10. (a) Dielectric constant and (b) loss factor versus temperature at several frequencies for pure PPO. Upturn at high temperature and low frequency is due to onset of conductance loss. Dashed line represents the extrapolated static dielectric constant. From Ref. 57, S. Yano, R. R. Rahalkar, S. P. Hunter, C. H. Wang, and R. H. Boyd, *J. Polym. Sci. Phys. Educ.*, Copyright (1976), John Wiley & Sons, Inc. Reprinted by permission of John Wiley & Sons, Inc., All Rights Reserved.

tributed mainly to the segmental motion of the polymer backbone. Below 100 MHz, the polymer relaxation falls off rapidly and thus PEO behaves as an insulator at low frequencies. Contrasting behavior of $NaBH_4$ and $NaSCN$ systems is observed: the former behaves very much as PEO, and the latter shows lower frequency-dependent conductivity arising from long-range ionic motion. The behavior of $NaBH_4$ systems may be explained by the particularly strong ion pairing that is thought to exist in these polymer electrolytes.[81] The dielectric constant of this system is little different

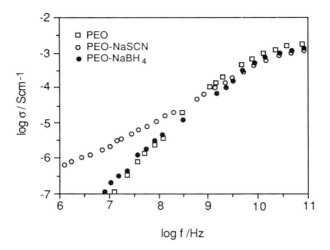

Figure 8.11. Frequency-dependent conductivity of PEO, PEO–NaSCN, and PEO–NaBH$_4$. From Ref. 76.

from that of the pure polymer, suggesting that the contribution to the polarizability is minimal, because ions either are unable to move to oppose the effect of the field or are very tightly bound to the polymer chains.

A study on PEO–NH$_4$CF$_3$SO$_3$ examined the effects of phase distribution. A metastable amorphous sample was prepared by fast quenching and was compared with equilibrated samples that showed approximately 70% crystallinity. Figure 8.12 shows the room temperature frequency-dependent conductivity. For the amorphous electrolyte conductivity shows no change up to 10 MHz but then increases rapidly. The crystalline complex behaves similarly to the amorphous one only above 1 MHz: at lower frequencies the conductivity drops. This raises the question of the nature of the ionic motion around 1 MHz in relation to high-energy barriers to long-range motion experienced at frequencies below 1 kHz. In addition, in contrast to the behavior of NaSCN systems described earlier, the conductivity of NH$_3$CF$_3$SO$_3$ electrolytes is seen to be higher at high frequency, implying that charge carriers contribute to the conductivity over and above that arising from polymer backbone motion.

The effect of salt concentration on the dielectric response has been studied in detail for a PEO–LiClO$_4$ electrolyte.[78] All samples were amorphous or could be maintained in a metastable amorphous state for the duration of the experiment. Figure 8.13 shows the frequency response of the permittivity ϵ' and loss ϵ'' for PEO and P(EO)$_{50}$LiClO$_4$, and Table 8.1 summarizes the dielectric data over the concentration range 100:1 to 6:1. All samples exhibited a dielectric dispersion in the gigahertz region of the frequency spectrum. This has been attributed to the β relaxation arising from long-range segmental motion of the polymer. For ion-coordinated systems, this relaxation is likely to involve cooperative motion of the ions

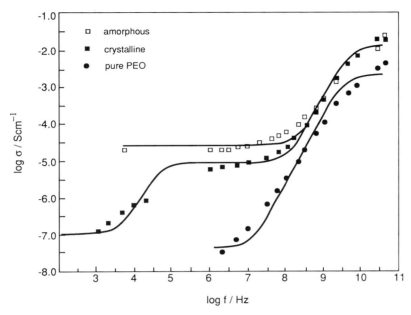

Figure 8.12. Room temperature conductivity/frequency plots for $P(EO)_8NH_4CF_3SO_3$. The lines represent a theoretical fit to experimental values. From Ref. 48.

with the polymer. The ions alter the average dipole moment and polarizability, as observed through permittivity variations, and hence the dispersion arises through a relaxation of coupled units. There are two contributions to the total measured absorption ϵ_T'',

$$\epsilon_T = \epsilon_D'' + (1.8 \times 10^{12} \, \sigma_{dc}/f)$$

where ϵ_D'' includes dipolar relaxation and short-range ionic vibrations and may be separated from σ_{dc}, the specific dc conductivity (S cm^{-1}). At the lower end of the frequency range, energy dissipation resulting from ionic conductance predominates. When the ionic conductivity was separated from the total loss, a second weak relaxation was observed in the megahertz region and was more clearly defined with increasing salt content. Although it is possible that this is an artifact of the subtraction process, it may well arise from the relaxation of ion aggregates, similar to that reported for ion-containing low-molecular-weight solvents.[82,83] This is backed by the behavior of the permittivity (Figure 8.13) which shows a continual increase to lower frequencies, which may result from a broad relaxation. The effects of adding salts to low- and high-dielectric-constant, low-molecular-weight solvents have indicated a broadening of the absorption curves with increased salt content.[83] This broadening has been associated with overlapping of additional relaxation processes arising from ion pairing. The dispersion amplitude

$$(\epsilon_0' - \epsilon_\infty') = 2\epsilon_m''/\beta$$

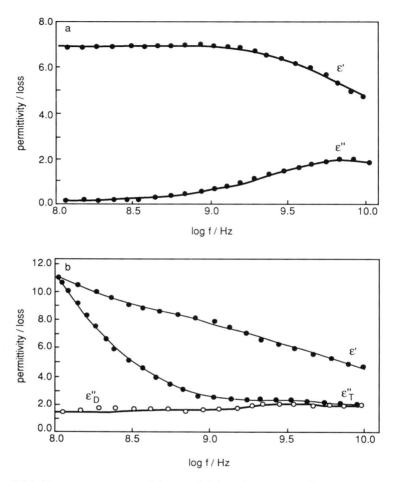

Figure 8.13. Frequency response of the permittivity (ϵ') and loss (ϵ'') for (**a**) PEO and (**b**) P(EO)$_{50}$LiClO$_4$. ϵ''_T is the total loss and ϵ''_D is the loss value after subtracting dc conductivity. From Ref. 78, F. M. Gray, C. A. Vincent, and M. Kent, *J. Polym. Sci. Phys. Educ.*, Copyright (1989), John Wiley & Sons, Inc. Reprinted by permission of John Wiley & Sons, Inc., All Rights Reserved.

where ϵ'_0 and ϵ'_∞ are the low- and high-frequency limiting values of the permittivity, ϵ''_m is the loss maximum, and β is the dispersion parameter with $0 < \beta \leq 1$, equal to unity for a single relaxation, is plotted against salt concentration in Figure 8.14 and shows a trend very similar to that of the conductivity variation with concentration. The increase in $\epsilon'_0 - \epsilon'_\infty$ at low salt content is likely to be related to the enhancement in conductivity through the increase in charge carrier concentration. At high salt content, the fall in the dispersion amplitude was interpreted as resulting from the greater restriction in segmental motion and ion mobility, changing the dipolar re-

Table 8.1. Dielectric Data for PEO and PEO–LiClO$_4$ Systems

0:Li	ϵ'_0	ϵ'_∞	ϵ''_m	f_m (GHz)	σ_{dc} (Ω^{-1} cm^{-1})	β
$T = 75°C$						
(PEO)	7.0 ± 0.3	3.5 ± 0.4	1.7 ± 0.2	6.7 ± 0.3	—	0.9
100:1	8.8 ± 0.4	3.7 ± 0.4	1.8 ± 0.2	4.7 ± 0.4	9.9×10^{-5}	0.7
50:1	9.6 ± 0.4	4.4 ± 0.6	1.9 ± 0.2	4.3 ± 0.4	3.5×10^{-4}	0.7
30:1	11.2 ± 0.4	4.7 ± 0.6	1.9 ± 0.2	3.5 ± 0.4	5.3×10^{-4}	0.7
20:1	13.5 ± 0.4	4.3 ± 0.6	2.0 ± 0.2	3.2 ± 0.5	6.0×10^{-4}	0.5
10:1	12.6 ± 0.4	4.5 ± 0.6	1.5 ± 0.2	2.7 ± 0.5	2.3×10^{-4}	0.4
6:1	8.8 ± 0.4	4.7 ± 0.6	0.9 ± 0.2	1.6 ± 0.5	6.6×10^{-5}	0.5
$T = 65°C$						
(PEO)	5.2 ± 0.3	3.1 ± 0.4	1.2 ± 0.2	6.3 ± 0.4	—	0.9
100:1	7.4 ± 0.4	3.5 ± 0.4	1.3 ± 0.2	3.6 ± 0.4	7.9×10^{-5}	0.7
50:1	8.5 ± 0.4	3.6 ± 0.4	1.9 ± 0.2	3.9 ± 0.5	2.3×10^{-4}	0.8
30:1	10.2 ± 0.4	4.4 ± 0.4	1.8 ± 0.2	3.3 ± 0.5	3.4×10^{-4}	0.7
20:1	10.9 ± 0.4	4.8 ± 0.6	1.9 ± 0.2	3.1 ± 0.5	4.0×10^{-4}	0.7
10:1	10.4 ± 0.4	4.8 ± 0.6	1.3 ± 0.2	2.5 ± 0.6	1.9×10^{-4}	0.6
6:1	7.8 ± 0.5	4.4 ± 0.6	0.8 ± 0.2	1.5 ± 0.6	5.3×10^{-5}	0.5
$T = 55°C$						
(PEO)	3.8 ± 0.3	2.6 ± 0.4	0.6 ± 0.2	3.3 ± 0.4	—	0.9
100:1	4.9 ± 0.4	2.8 ± 0.4	0.6 ± 0.2	3.2 ± 0.4	—	0.6
50:1	8.6 ± 0.4	3.8 ± 0.4	1.8 ± 0.2	3.2 ± 0.6	1.6×10^{-4}	0.7
30:1	9.5 ± 0.4	4.3 ± 0.6	1.7 ± 0.2	3.6 ± 0.6	1.5×10^{-4}	0.7
20:1	9.6 ± 0.4	4.1 ± 0.5	1.7 ± 0.2	2.9 ± 0.6	2.5×10^{-4}	0.7
10:1	9.1 ± 0.5	4.5 ± 0.6	1.0 ± 0.2	2.3 ± 0.6	9.5×10^{-5}	0.5
6:1	7.2 ± 0.5	4.2 ± 0.6	0.6 ± 0.2	1.5 ± 0.7	2.6×10^{-5}	0.4

sponse to the external field. The effects of widespread ion aggregation must contribute to the increase in the dispersion amplitude but whether the high concentration decrease reflects a possible breakup of aggregates, as a result of increased polarizability of the matrix or a separation of clusters that are less affected by the polymer response, or is predominately the result of mobility restrictions is not clear at this stage.

Audio frequency dielectric measurements have been undertaken by Fontanella, Wintersgill, and co-workers[71,84–91] on PEO– and PPO–salt systems. Measurements have been made as a function of temperature at discrete frequencies. The temperature and frequency range employed enabled studies of the γ relaxation to be carried out. However, the β relaxation associated with the glass transition and observed in the high-frequency studies described earlier was often unobserved because the T_g of the salt-containing polymer is sufficiently moved to higher temperatures for it to be masked by effects of high background conductivity.

The γ relaxation for PEO containing LiClO$_4$, LiSCN, and NaSCN has been reported to be very similar to that of pure PEO. This is surprising as it may be expected that high concentrations of salt would significantly affect the short-range mo-

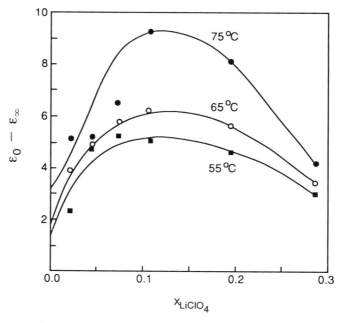

Figure 8.14. Variation of the dispersion amplitude ($\epsilon'_0 - \epsilon'_\infty$) with $LiClO_4$ concentration in PEO. From Ref. 78, F. M. Gray, C. A. Vincent, and M. Kent, *J. Polym. Sci. Phys. Educ.*, Copyright (1989). John Wiley & Sons, Inc. Reprinted by permission of John Wiley & Sons, Inc., All Rights Reserved.

tion of the polymer segments. Fontanella and co-workers[90,91] suggested that a $tg^+t \leftrightarrow tg^-t$ mechanism[92] in contracted PEO helices might be relatively unaffected by the presence of small ions. This would be consistent with results found for KSCN complexes. Three peaks were resolved in this instance, indicating that distortions resulting from cation size had been introduced that produced inequivalences in the barriers associated with the polymer motion. Likewise, the $NaClO_4$[91] system exhibited multiple peaks in this temperature region, with the relative peak heights concentration dependent. By comparison with the behavior of PEO–NaSCN electrolytes, it would appear that the short-range chain motion was dependent on both cation and anion. An extra low-temperature absorption occurs in ClO_4^- samples and has been attributed to perchlorate-induced defects on the exterior of the polymer chain. Alkaline earth salts $Ca(SCN)_2$ and $Ba(SCN)_2$ produce shifts in the γ relaxation more pronounced for the barium salt, where the relaxation moves to higher temperature and hence higher activation energy.[84] Interestingly, PEO complexed with this salt exhibits the higher conductivity.

A number of studies[85,93] on amorphous polymer–salt systems have demonstrated that the electrical conductivity has temperature and pressure dependencies very similar to those of the β relaxation. This is discussed in more detail in Chapter 5.

The findings substantiate the idea that the ionic conductivity is dominated by large-scale segmental motion in the host polymer.

Relaxation studies carried out using TSDC[72] indicate that the γ relaxation contains several closely spaced relaxations and it is the relative populations of these that change, giving rise to shifts in the relaxation spectrum. Figure 8.15 shows the TSDC spectra for PEO with NaClO$_4$ and NaSCN complexes of PEO. Although the dielectric spectrum indicates one broad peak[90,91] three closely spaced peaks are observed by this method.

Measurements of the temperature dependence of the dielectric relaxation times of PPO network polymer electrolytes containing LiClO$_4$ have been compared with thermal *cis–trans* isomerization rates of azobenzene chromophores incorporated as crosslinks into the network.[94] Relaxation times and conductivity correlated well and indicated that ionic conductivity was closely associated with chain segmental motion. In contrast, the thermal isomerization rate was hardly affected as the T_g increased with increasing salt concentration and crosslink density. Similar studies were carried out by Greenbaum[95] and Cameron et al.,[96] who determined the correlation times of a tumbling paramagnetic probe molecule from electron paramagnetic resonance (EPR) spectra of P(EO)$_{4.5}$NaClO$_4$ and a low-molecular-weight EO–PO copolymer containing LiClO$_4$. The probe molecule was not bonded to the polymer in this instance so rotational behavior was not directly related to the polymer motion. In the former study, a distinct fall in activation energy for the process was noted above 120°C and could be related to the increase in chain mobility. Differential scanning calorimetry traces show a sharp thermal event around 150°C and suggest melting of a crystalline complex. For the latter study, addition of LiClO$_4$ to the polymer progressively reduced the mobility of the probe

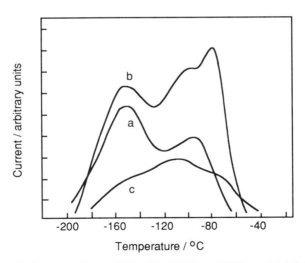

Figure 8.15. TSDC spectra for (**a**) PEO, (**b**) P(EO)$_{4.5}$NaSCN, and (**c**) P(EO)$_{4.5}$NaClO$_4$. From Ref. 72.

molecule, although the polymer molecular weight had little effect. Reduction in mobility was not a result, therefore, of increased macroscopic viscosity, but was a response to microscopic changes in the local environment. The effective volume of the relaxing polymer segment decreased with increasing salt content. This may be envisaged as transient crosslinking by the ions, effectively reducing the segmental size.

8.3.3. Nuclear Magnetic Resonance Spectroscopy

NMR spectroscopy may be used as a probe of both molecular environment and dynamics. Experimentally, NMR yields values of the motional correlation time τ_c that can be related to the diffusion coefficient and, for translational diffusion, can be interpreted as the mean time an atom resides on a particular site. An extremely wide range of diffusion coefficient magnitudes can be explored by the various NMR techniques. Also, NMR methods have a particular advantage in probing dynamics in specific nuclei. The techniques and theories relating to NMR are covered in numerous texts,[97,98] and the principles of the techniques that are applicable to polymer electrolytes have been reviewed by Chadwick and Worboys.[99]

The majority of experiments relating to polymer electrolytes have involved either linewidth or relaxation time studies. For the former, radiation in the form of an oscillating magnetic field of strength B_1 is applied perpendicular to the static magnetic field B_0. The radiation is swept and resonance detected by an absorption of energy. Relative motion of the nuclei caused by translational diffusion will have the effect of averaging out the local magnetic field and the observed resonance line will narrow. "Motional line narrowing" occurs when τ_c is typically less than 10^{-4} s. Studies of linewidth as a function of temperature provide a simple way of detecting diffusion effects. Pulsed experiments require the sample to be subjected to short-time high-intensity radio frequency energy which disturbs the spin populations in various energy levels. Often, complex pulse sequences are employed. The return to equilibrium conditions is monitored via the magnetization along particular directions. The relaxation of spins requires that they interact with magnetic fields oscillating at a frequency corresponding to the separation between energy levels. The motion of neighboring spins will create an oscillating dipolar field at a given lattice site with a wide frequency distribution and can cause relaxation. The relaxation time for an ensemble of spins depends on the intensity of the component in the frequency distribution of the dipolar field at the value required to give relaxation. The commonly measured relaxation times are T_1, the spin–lattice relaxation time; T_2, the spin–spin relaxation time, $T_{1\rho}$, the spin–lattice relaxation time in the rotating frame; and T_{1D}, the spin–lattice relaxation time in the local dipolar field. These measurements have been used to obtain information on the local environment of principally lithium and sodium nuclei in polymers. Studies of diffusional processes by these two techniques have not been very successful mainly because of problems in relating NMR parameters to long-range diffusion as distinct from short-range motional disorder. Diffusion coefficients can be measured directly using pulsed field gradient NMR, a method similar to radiotracer techniques where a nuclear spin is used as a label on the atom. The use of these techniques to measure diffusion coefficients is considered in Chapter 9.

The phase behavior of PEO–LiCF$_3$SO$_3$, PEO–LiClO$_4$, and PEO–NaI has been studied by Berthier and co-workers,[100–102] who assigned nuclei to crystalline or amorphous phases on the basis of the decay time of the transverse nuclear magnetization correlation function T_2. In Figure 8.16 the fractions of protons and fluorines in the crystalline phase were shown as a function of temperature for PEO–LiCF$_3$SO$_3$. From these studies arose the important conclusion that the amorphous phase was responsible for the ionic conductivity. Tunstall et al.[103,104] also measured relaxations for ^{7}Li, ^{19}F, and ^{1}H. These studies showed that the T_2 signal could not simply be partitioned between a crystalline phase and an elastomeric phase. The difference in behavior of the ^{1}H and ^{19}F signals was thought to be due to a contribution to the proton signal from the crystalline fraction arising from some PEO chains undergoing segmental relaxations at sufficiently high frequency to motionally narrow the spectrum. Donoso et al.[105] have measured ^{1}H, ^{7}Li, and ^{19}F NMR relaxation times in the amorphous phase of P(EO)$_8$LiBF$_4$. Results suggest that cation motion is controlled by segmental motion of the polymer chain, but the BF$_4^-$ anion has an additional degree of freedom associated with BF$_4$ rotation.

^{7}Li linewidth and T_1 measurements have been used to study the PVA–LiClO$_4$ system.[106–108] A single absorption line with no quadrupolar broadening was reported with motional narrowing similar to that found for PEO-based systems. The T_1 data for an 8:1 material could be fitted to an Arrhenius plot with a very low activation energy (14 kJ mol^{-1}). This was associated with local motions that were

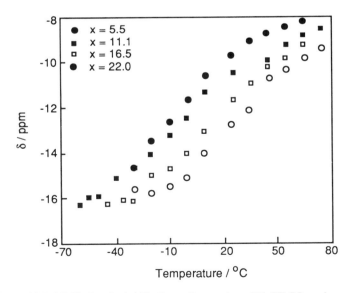

Figure 8.16. ^{23}Na NMR chemical shifts for a siloxane-based NaCF$_3$SO$_3$ polymer electrolyte as a function of salt concentration and temperature. From Ref. 117. Reprinted with permission from *J. Am. Chem. Soc.* **110** (1988), 3036. Copyright 1988 American Chemical Society.

not necessarily related to long-range ionic transport. In 4:1 samples two separate processes were identified as contributing to the relaxation.

Greenbaum[95] studied ^{23}Na and ^{1}H NMR in PEO–NaClO$_4$. The ^{23}Na line at a resonance frequency of 105.8 MHz showed no significant quadrupolar broadening, suggesting a symmetric environment for the ions. As in lithium-based systems, motional narrowing occurred but in this system it was detectable at temperatures as low as $-50°C$, and the activation energy for this process is considerably less than that for ionic conductivity.[99] In a similar study of the PEO–NaSCN system, Chadwick and Worboys[99] noted that at low temperatures, the sodium line was broad and exhibited a structure that could be ascribed to sodium ions in nonequivalent sites, possibly associated with crystalline and amorphous phases. For PPO–NaSCN only a simple line was observed at all temperatures. Greenbaum and co-workers[93,109–112] have reported studies of amorphous high-molecular-weight PPO and crosslinked siloxane–polyether-based electrolytes containing a number of sodium salts. ^{23}Na-free induction decay spin–lattice signals contained two readily distinguishable components. The short T_1 component, which was found to narrow above T_g, was attributed to "mobile" Na$^+$, whereas the long T_1 resonance was not noticeably temperature dependent until relatively high temperatures when broadening was observed. The latter signal was related to some associated species. ^{23}Na NMR studies on MEEP–NaCF$_3$SO$_3$ show results similar to those for the other amorphous ion-conducting polymers.[113] Ion aggregation and salt precipitation trends in various materials were followed by measuring mobile/bound Na$^+$ concentrations. The ^{23}Na chemical shifts of the mobile Na$^+$ resonance exhibit significant dependencies on both anion and temperature, indicating that cation–anion association plays a role in the transport mechanism. The presence of specific configurations such as ion pairs cannot be established; only information about average ion association effects may be deduced. The effect of applying high pressures (up to 0.2 GPa) to samples has been followed by ^{23}Na NMR.[93,110,112] Conductivity in PPO–salt complexes is known to decrease as pressure is increased.[84,93] Increased pressure was found to broaden the resonance. These phenomena may be indicative of subtle ion–polymer interactions or may simply reflect the pressure dependence of the T_g.

Donoso et al.[114–116] have reported ^{1}H and ^{31}P relaxation time data for PEO–H$_3$PO$_4$ and PEO–D$_3$PO$_4$ proton conductors. The temperature dependence of the conductivity and the proton correlation time were found to be very similar, showing the proton diffusion in PEO–(H$_3$PO$_4$)$_{0.42}$ to be governed by chain segmental motion. The temperature dependencies of ^{31}P relaxation rates for protonated and deuterated acid show a peak in $T_{1\rho}{}^{-1}$ around 260 K with the same amplitude. Its origin has been attributed to a modulation of the interaction between protons in the chain and the phosphorus nuclei, ruling out intramolecular dipole–dipole interactions as the dominant relaxation process. At high temperature, a new relaxation mechanism was observed in the ^{31}P spectra which may be the result of rotational–translational motion of the molecule.

High-resolution magic-angle-spinning (MAS) solid-state NMR is a potentially powerful probe for polymer electrolytes. One study to date has been reported by Spindler and Shriver[117] in which ^{13}C, ^{29}Si, ^{7}Li, and ^{23}Na solid state-NMR characterization of an amorphous siloxane-based polymer electrolyte was undertaken. Lithium and sodium trifluoromethane sulfonate complexes show different chemical shift behavior: the former was invariant at -1.3 ppm with salt concentration and temperature; for the latter, the resonances shifted upfield as salt concentration increased and temperature decreased, approaching limiting values at the upper and lower ends of the temperature range (-70 to $80°C$). These data suggest an increasing degree of anion–cation interaction with increasing salt content. The ^{7}Li linewidths were sensitive to concentration and temperature, broadening at lower temperature. ^{13}C linewidths and T_1 values, along with ^{7}Li linewidths, correlated with the T_g. Both ^{13}C chemical shifts and T_1 values indicate that the ether carbons closest to the siloxane backbone are not involved in coordination of the lithium cations.

8.3.4. Quasi-elastic Neutron Scattering

QENS may be used to obtain information about rotational and translational motions of protons on an atomic scale. When a neutron interacts with a nucleus, it can be either absorbed or scattered. For simplicity, scattering will be considered as the only neutron–nucleus interaction. The incident neutron beam is of a fixed wavelength, with constant energy E_0 and wave vector $\mathbf{k}_o$. The momentum of the incident neutrons is given by $\hbar\mathbf{k}_o$ and therefore

$$E_0 = \frac{\hbar^2 k_0^2}{2m} = \hbar\omega_0$$

where m is the mass of the neutron. After scattering from the sample, a neutron will possess an energy E and a new wave vector $\mathbf{k}$, with

$$E = \frac{\hbar^2 \mathbf{k}^2}{2m} = \hbar\omega$$

Thus in a QENS experiment, the energy change and wave vector change, Q ($= \mathbf{k} - \mathbf{k}_o$), must be measured. Because the scattering is quasi-elastic, the energy change is effectively zero. Likewise, the size of the wave vector will be little changed, but after scattering, the neutron may be moved through a solid angle Ω, giving a change in scattering cross section, σ, with Ω and E with

$$\frac{\partial^2\sigma}{\partial\Omega\partial E} = 1/\hbar \ \frac{\partial^2\sigma}{\partial\Omega\partial\omega}$$

and this in turn is represented in terms of a scattering law, $S(Q,\omega)$. Neutron scattering may be coherent or incoherent depending on whether phase changes between the scattered and incident neutrons are introduced. In the case of hydrogen, the scattering law is dominated by the incoherent contribution to the cross section. For deuterium, the coherent contribution to the cross section is more important and therefore partial deuteration of samples for neutron diffraction experiments changes the cross section sufficiently to allow the position of particular hydrogen nuclei in a structure to be determined. A much more detailed description of this technique can be found in Ref. 118 and a review by Poinsignon[119] highlights the application of QENS to studies of proton transfer in solid protonic conductors.

Few QENS studies have as yet been carried out on polymer electrolytes. Donoso et al.[115,116] have reported results for studies on PEO–H_3PO_4 and PEO–D_3PO_4. The fact that only small differences between measurements for H_3PO_4 and D_3PO_4 containing polymers exist indicates that motion of protons belonging to the chain and motion of protons belonging to the acid are closely related. A study by Hardgrave[120] on PEO–$LiCF_3SO_3$ systems focused on the motion of the methylene groups in the polymer chains and correlated this with conductivity urements over the same salt concentration range. As may be expected, the displacement volume of the methylene groups was found to contract as the salt concentration increased and followed the decrease in conductivity of these systems.

References

1. M. L. Williams, R. F. Landel, and J. D. Ferry, *J. Am. Chem. Soc.* **77** (1955), 3701.

2. M. Ratner, in *Polymer Electrolyte Reviews—1* (J. R. MacCallum and C. A. Vincent, Eds.), Elsevier Applied Science, London (1987), p. 173.

3. H. Cheradame and J. F. Le Nest, in *Polymer Electrolyte Reviews—1* (J. R. MacCallum and C. A. Vincent, Eds.), Elsevier Applied Science, London (1987), p. 103.

4. A. Killis, J. F. Le Nest, A. Gandini, and H. Cheradame, *Macromolecules* **17** (1984), 63.

5. A. Killis, J. F. Le Nest, and H. Cheradame, *Makromol. Chem. Rapid Commun.* **1** (1980), 595.

6. A. Killis, J. F. Le Nest, A. Gandini, and H. Cheradame, *J. Polym. Sci. Polym. Phys. Ed.* **19** (1981), 1073.

7. H. Cheradame, in *IUPAC Macromolecules* (H. Benoit and P. Rempp, Eds.), Pergamon Press, Oxford (1982).

8. A. Killis, J. F. Le Nest, H. Cheradame, and A. Gandini, *Makromol. Chem.* **183** (1982), 2835.

9. A. Killis, J. F. Le Nest, H. Cheradame, and A. Gandini, *Makromol. Chem.* **183** (1982), 1037.

10. M. H. Cohen and D. Turnbull, *J. Chem. Phys.* **31** (1959), 1164.

11. H. Vogel, *Phys. Z.* **22** (1921), 645; G. Tamman and W. Hesse, *Z. Anorg. Allg. Chem.* **156** (1926), 245; G. S. Fulcher, *J. Am. Ceram. Soc.* **8** (1925), 339.

12. R. Syed, D. L. Gavin, and C. T. Moynihan, *J. Am. Ceram. Soc.* **65** (1982), C129.

13. C. S. Harris, D. F. Shriver, and M. A. Ratner, *Macromolecules* **19** (1986), 987.

14. N. Kobayashi, M. Uchiyama, K. Shigehara, and E. Tsuchida, *J. Phys. Chem.* **89** (1985), 987.

15. J. E. Weston and B. C. H. Steele, *Solid State Ionics* **2** (1981), 347.

16. A. J. Batschinsky, *Z. Phys. Chem.* **84** (1913), 644.

17. T. Miyamoto and K. Shibayama, *J. Appl. Phys.* **44** (1973), 5372.

18. M. Watanabe and N. Ogata, in *Polymer Electrolyte Reviews—1* (J. R. MacCallum and C. A. Vincent, Eds.), Elsevier Applied Science, London (1987), p. 39.

19. M. Watanabe, J. Ikeda, and I. Shinohara, *Polym. J.* **15** (1983), 65.

20. M. Watanabe, J. Ikeda, and I. Shinohara, *Polym. J.* **15** (1983), 175.

21. M. Watanabe, K. Sanui, N. Ogata, T. Kobayashi, and Z. Ohtaki, *J. Appl. Phys.* **57** (1985), 123.

22. M. Watanabe, K. Sanui, and N. Ogata, *Macromolecules* **19** (1986), 815.

23. M. Watanabe, S. Oohashi, K. Sanui, N. Ogata, T. Kobayashi, and Z. Ohataki, *Macromolecules* **18** (1985), 1945.

24. J. H. Gibbs and E. A. di Marzio, *J. Chem. Phys.* **28** (1958), 373.

25. G. Adams and J. H. Gibbs, *J. Chem. Phys.* **43** (1965), 139.

26. B. L. Papke, M. A. Ratner, and D. F. Shriver, *J. Electrochem. Soc.* **129** (1982), 1694.

27. D. F. Shriver, R. Dupon, and M. Stainer, *J. Power Sources* **9** (1983), 383.

28. C. A. Angell, *Solid State Ionics* **9/10** (1983), 3.

29. C. A. Angell and W. Sichina, *Ann. N.Y. Acad. Sci.* **279** (1976), 53.

30. J. M. G. Cowie, A. C. S. Martin, and A. M. Firth, *Br. Polym. J.* **20** (1988), 247.

31. C. A. Angell, *Solid State Ionics* **18/19** (1986), 72.

32. J. J. Fontanella, M. C. Wintersgill, J. P. Calame, M. K. Smith, and C. G. Andeen, *Solid State Ionics* **18/19** (1986), 253.

33. M. C. Wintersgill, J. J. Fontanella, M. K. Smith, S. G. Greenbaum, K. J. Adamic, and C. G. Andeen, *Polymer* **28** (1987), 633.

34. J. R. M. Giles, *Solid State Ionics* **24** (1987), 155.

35. C. A. Angell and R. D. Bressel, *J. Phys. Chem.* **76** (1972), 3244.

36. S. G. Greenbaum, K. J. Adamic, Y. S. Pak, M. C. Wintersgill, and J. J. Fontanella, *Solid State Ionics* **28–30** (1988), 1042.

37. N. Kobayashi, M. Uchiyama, K. Shigehara, and E. Tsuchida, *J. Phys. Chem.* **89** (1985), 987.

38. W. T. Laughlin and D. R. Uhlmann, *J. Phys. Chem.* **76** (1972), 2317.

39. C. A. Angell, *Solid State Ionics* **9/10** (1983), 3.

40. C. A. Angell, *Solid State Ionics* **18/19** (1986), 72.

41. L. M. Torrell and C. A. Angell, *Br. Polym. J.* **20** (1988), 173.

42. S. D. Druger, M. A. Ratner, and A. Nitzan, *Solid State Ionics* **9/10** (1983), 1115.

43. S. D. Druger, M. A. Ratner, and A. Nitzan, *J. Chem. Phys.* **79** (1983), 3133.

44. S. D. Druger, M. A. Ratner, and A. Nitzan, *Phys. Rev. B* **31** (1985), 3939.

45. S. D. Druger, in *Transport and Relaxation Processes in Random Materials* (J. Klafter, R. J. Rubin, and M. F. Shlesinger, Eds.), World Scientific, Singapore (1986).

46. C. S. Harris, A. Nitzan, M. A. Ratner, and D. F. Shriver, *Solid State Ionics* **18/19** (1986), 151.

47. S. D. Druger, M. A. Ratner, and A. Nitzan, *Solid State Ionics* **18/19** (1986), 106.

48. S. M. Ansari, M. Brodwin, M. Stainer, S. D. Druger, M. A. Ratner, and D. F. Shriver, *Solid State Ionics* **17** (1985), 101.

49. R. Granek, A. Nitzan, S. D. Druger, and M. A. Ratner, *Solid State Ionics* **28–30** (1988), 120.

50. M. A. Ratner and A. Nitzan, *Solid State Ionics* **28–30** (1988), 3.

51. S. D. Druger and M. A. Ratner, *Chem. Phys. Rev. B* **38** (1988), 1258.

52. S. D. Druger and M. A. Ratner, *Chem. Phys. Lett.* **151** (1989), 434.

53. M. A. Ratner and A. Nitzan, *Faraday Discuss. Chem. Soc.* **88** (1989), 19.

54. L. M. Torell and S. Schantz, in *Polymer Electrolyte Reviews—2* (J. R. MacCallum and C. A. Vincent, Eds.), Elsevier Applied Science, London (1989), p. 1.

55. H. Liu, Y. Okamodo, T. Skotheim, Y. S. Pak, S. G. Greenbaum, and K. J. Adamic, in *Solid State Ionics,* Vol. **135** (G. Nazri, R. A. Huggins, and D. F. Shriver, Eds.), Materials Research Soc., Pittsburgh (1989), p. 337.

56. M. C. Wintersgill, J. J. Fontanella, S. G. Greenbaum, D. Teeters, and R. Frech, *Solid State Ionics* **18/19** (1986), 271.

57. S. Yano, R. R. Rahalkar, S. P. Hunter, C. H. Wang, and R. H. Boyd, *J. Polym. Sci. Polym. Phys. Ed.* **14** (1976), 1877.

58. J. Sandahl, S. Schantz, L. M. Torell, and R. Frech, *Solid State Ionics* **28–30** (1988), 958.

59. J. Sandahl, S. Schantz, L. Börjesson, L. M. Torell, and J. R. Stevens, *J. Chem. Phys.* **91** (1989), 655.

60. S. Schantz, L. M. Torell, and J. R. Stevens, *J. Appl. Phys.* **64** (1988), 2038.

61. M. Watanabe, J. Ikeda, and I. Shinohara, *Polym. J.* **15** (1983), 65.

62. M. Watanabe, K. Sanui, N. Ogata, F. Inoue, T. Kobayashi, and Z. Ohtaki, *Polym. J.* **17** (1985), 549.

63. M. Watanabe, M. Itoh, K. Sanui, and N. Ogata, *Macromolecules* **20** (1987), 569.

64. G. G. Cameron, J. L. Harvie, M. D. Ingram, and G. A. Sorrie, *Br. Polym. J.* **20** (1988), 199.

65. G. Williams and D. C. Watts, *Trans. Faraday Soc.* **66** (1970), 80.

66. L. Börjesson, J. R. Stevens, and L. M. Torell, *Polymer* **28** (1987), 1803.

67. J. R. Stevens and S. Schantz, *Polym. Commun.* **29** (1988), 330.

68. C. H. Wang, G. Fytas, D. Lilge, and Th. Dorfmüller, *Macromolecules* **14** (1981), 1363.

69. C. H. Porter and R. H. Boyd, *Macromolecules* **4** (1971), 589.

70. T. M. Connor, B. E. Read, and G. Williams, *J. Appl. Chem.* **14** (1964), 74.

71. M. C. Wintersgill and J. J. Fontanella, in *Polymer Electrolyte Reviews—2* (J. R. MacCallum and C. A. Vincent, Eds.), Elsevier Applied Science, London (1989), p. 43.

72. D. R. Figueroa, J. J. Fontanella, M. C. Wintersgill, J. P. Calame, and C. G. Andeen, *Solid State Ionics* **28–30** (1988), 1023.

73. N. G. McCrum, B. E. Read, and G. Williams, *Anelastic and Dielectric Effects in Polymeric Solids*, John Wiley and Sons, New York (1967).

74. Y. Ishida, *J. Polym. Sci. A2* **7** (1969), 1835.

75. M. E. Bauer and W. H. Stockmayer, *J. Chem. Phys.* **43** (1965), 4319.

76. T. Wong, M. Brodwin, B. L. Papke, and D. F. Shriver, *Solid State Ionics* **5** (1981), 689.

77. F. M. Gray, C. A. Vincent, and M. Kent, *Solid State Ionics* **28–30** (1988), 936.

78. F. M. Gray, C. A. Vincent, and M. Kent, *J. Polym. Sci. Polym. Phys. Educ.* **27** (1989), 2011.

79. S. M. Ansari, M. Brodwin, M. Stainer, S. D. Druger, M. A. Ratner, and D. F. Shriver, *Solid State Ionics* **17** (1985), 101.

80. T. Wong, M. Brodwin, J. I. McOmber, and D. F. Shriver, *Solid State Commun.* **35** (1980), 591.

81. B. L. Papke, R. Dupon, M. A. Ratner, and D. F. Shriver, *Solid State Ionics* **5** (1981), 685.

82. N. E. Hill, in *Dielectric Properties and Molecular Behaviour* (N. E. Hill, W. E. Vaughan, A. H. Price, and M. Davies, Eds.), Van Nostrand, London (1969).

83. S. Onishi, H. Farber, and S. Petrucci, *J. Phys. Chem.* **84** (1980), 2922.

84. J. J. Fontanella, M. C. Wintersgill, J. P. Calame, and C. G. Andeen, *J. Polym. Sci. Polym. Phys. Ed.* **23** (1985), 113.

85. J. J. Fontanella, M. C. Wintersgill, M. K. Smith, J. Semancik, and C. G. Andeen, *J. Appl. Phys.* **60** (1986), 2665.

86. J. J. Fontanella, M. C. Wintersgill, M. K. Smith, S. G. Greenbaum, K. J. Adamic, and C. G. Andeen, *Polymer* **28** (1987), 633.

87. J. J. Fontanella, M. C. Wintersgill, P. J. Welcher, and C. G. Andeen, *J. Appl. Phys.* **58** (1985), 2875.

88. M. C. Wintersgill, J. J. Fontanella, J. P. Calame, D. R. Figueroa, and C. G. Andeen, *Solid State Ionics* **11** (1983), 151.

89. J. J. Fontanella, M. C. Wintersgill, J. P. Calame, M. K. Smith, and C. G. Andeen, *Solid State Ionics* **18/19** (1986), 253.

90. J. J. Fontanella, M. C. Wintersgill, J. P. Calame, and C. G. Andeen, *Solid State Ionics* **8** (1983), 333.

91. M. C. Wintersgill, J. J. Fontanella, J. P. Calame, D. R. Figueroa, and C. G. Andeen, *Solid State Ionics* **11** (1983), 151.

92. T. Suzuki and T. Kotaka, *Macromolecules* **13** (1980), 1495.

93. S. G. Greenbaum, Y. S. Pak, M. C. Wintersgill, J. J. Fontanella, J. W. Schultz, and C. G. Andeen, *J. Electrochem. Soc.* **135** (1988), 235.

94. M. Watanabe, A. Suzuki, T. Sano, K. Sanui, and N. Ogata, *Macromolecules* **19** (1986), 1921.

95. S. G. Greenbaum, *Solid State Ionics* **15** (1985), 259.

96. G. G. Cameron, M. D. Ingram, B. Munro, and E. Ross, *Eur. Polym. J.* **24** (1988), 395.

97. A. Abragam, *The Principles of Nuclear Magnetism,* Oxford University Press, Oxford (1961).

98. C. P. Slichter, *The Principles of Magnetic Resonance,* Harper and Row, New York (1978).

99. A. V. Chadwick and M. R. Worboys, in *Polymer Electrolyte Reviews—1* (J. R. MacCallum and C. A. Vincent, Eds.), Elsevier Applied Science, London (1987), p. 275.

100. C. Berthier, W. Gorecki, M. Minier, M. B. Armand, J. M. Chabagno, and P. Rigaud, *Solid State Ionics* **11** (1983), 91.

101. M. Minier, C. Berthier, and W. Gorecki, *J. Phys.* **45** (1984), 739.

102. W. Gorecki, R. Andreani, C. Berthier, M. B. Armand, M. Mali, J. Roos, and D. Brinkmann, *Solid State Ionics* **18/19** (1986), 295.

103. D. P. Tunstall, A. S. Tomlin, J. R. MacCallum, and C. A. Vincent, *J. Phys. C.* **21** (1988), 1039.

104. D. P. Tunstall, A. S. Tomlin, F. M. Gray, J. R. MacCallum, and C. A. Vincent, *J. Phys. Condens. Matter* **1** (1989), 4035.

105. J. P. Donoso, M. G. Cavalcante, W. Gorecki, C. Berthier, and M. Armand, *Proceedings, XXV Congress Ampère on Magnetic Resonance, Stuttgart, September* (1990).

106. M. C. Wintersgill, J. J. Fontanella, J. P. Calame, S. G. Greenbaum, and C. G. Andeen, *J. Electrochem. Soc.* **131** (1984), 2208.

107. M. C. Wintersgill, J. J. Fontanella, J. P. Calame, S. G. Greenbaum, K. J. Adamic, A. N. Shetty, and C. G. Andeen, *Solid State Ionics* **18/19** (1986), 326.

108. S. G. Greenbaum and J. J. Fontanella, in *Relaxation in Complex Systems* (K. L. Ngai and A. G. B. Wright, Eds.), NTIS, Springfield, Va. (1984).

109. K. J. Adamic, S. G. Greenbaum, M. C. Wintersgill, and J. J. Fontanella, *J. Appl. Phys.* **60** (1986), 1342.

110. M. C. Wintersgill, J. J. Fontanella, S. G. Greenbaum, and K. J. Adamic, *Br. Polym. J.* **20** (1988), 195.

111. S. G. Greenbaum, Y. S. Pak, M. C. Wintersgill, and J. J. Fontanella, *Solid State Ionics* **31** (1988), 241.

112. S. G. Greenbaum, K. J. Adamic, Y. S. Pak, M. C. Wintersgill, J. J. Fontanella, D. A. Beam, and C. G. Andeen, in *Proceedings, Electrochemical Society Symposium on Electro-Ceramics and Solid State Ionics* (H. Tuller, Ed.), Pennington, N.J. (1988).

113. S. G. Greenbaum, K. J. Adamic, Y. S. Pak, M. C. Wintersgill, and J. J. Fontanella, *Solid State Ionics* **28–30** (1988), 1042.

114. P. Donoso, W. Gorecki, C. Berthier, F. Defendini, and M. B. Armand, *First International Symposium on Polymer Electrolytes (ISPE-1), St. Andrews* (1987), paper 17.

115. P. Donoso, W. Gorecki, C. Berthier, F. Defendini, C. Poinsignon, and M. B. Armand, *Solid State Ionics* **28–30** (1988), 969.

116. P. Donoso, W. Gorecki, C. Berthier, F. Defendini, C. Poinsignon, and M. B. Armand, in *Polymer Motion in Dense Systems* (D. Richter and T. Springer, Eds.), Springer-Verlag, Berlin (1988).

117. R. Spindler and D. F. Shriver, *J. Am. Chem. Soc.* **110** (1988), 3036.

118. M. Bée, *Ap lication of Quasielastic Neutron Scattering to Solid State Chemistry, Biology and Polymers,* Adam Helger, Bristol (1988).

119. C. Poinsignon, *Solid State Ionics* **35** (1989), 107.

120. M. D. Hardgrave, Ph.D. thesis, University of St. Andrews (1990).

CHAPTER 9

Transport Properties: Ionic Species and Mobility

It has been established that many salts dissolve in high-molecular-weight polyethers to form ionic conductors and that the solvent, although immobile in the macroscopic sense, plays a very important role in the conduction process through local chain flexibility. Many questions relating to the type of species involved in charge transport remain unanswered, however.

9.1. Ion–Ion Interactions

When a salt is dissolved in a polymer matrix, the conductivity, σ, increases as a result of the addition of charge carriers; however, as the salt concentration is increased above approximately 0.1 mol dm^{-3} (O:M ratio of $\sim$100:1 to 50:1) the conductivity reaches a maximum and then falls. This is thought to result from the introduction of an ever-increasing number of transient crosslinks in the system, reducing chain mobility. T_g measurements and dielectric studies do indeed support this view and have been discussed in detail in Chapter 8. It has also been suggested that at high salt concentrations, formation of immobile aggregated species will contribute to the fall in conductivity. It is in fact found that for conductivity data correlated to salt content at a reduced temperature $(T - T_g)$; that is, at constant mobility, conductivity maxima are still observed.[1-7] If these species do indeed exist, they are likely to have retarded diffusion rates[8]; however, as implied by Watanabe, it is difficult to distinguish between the effects of ion association and long-range interionic interactions on the reduced ionic mobility.

In an ionic solution, the proximity of ions subjects each to an electric field that influences its behavior, and competition exists between thermal forces that try to randomize the ionic distribution and electrostatic forces that attempt to order the system to minimize potential energy. The concept of the ion pair was introduced by Bjerrum in 1926. In this theory, when the separation distance of two oppositely charged ions falls below a somewhat arbitrary minimum value, where the ions' attractive electrostatic potential energy is greater than that of thermal motion, then the ions are not considered as separate species and the resulting ion pair is assumed to be unaffected by an external field. This definition of association is limited to high dilutions. As the salt concentration in a solution is increased, the interionic distance decreases and ion–ion interactions become progressively more significant. Fuoss[9] made a more detailed analysis of the formation of ion pairs and showed that above a critical concentration, interactions of higher order become significant and pairwise coulombic attraction no longer serves as a basis for distinguishing between free and associated ions because of space crowding. Pettit and Bruckenstein[10] showed later that in very low dielectric constant media, clusters of three, four, and six ions may become thermally stable. Ion pairs may be defined as contact or solvent-separated species. The former is produced when a molecule of the first-neighbor solvent shell around an ion is replaced by an oppositely charged ion. The latter results when an overlap of solvent cospheres occurs and links a pair of ions to form a mechanical unit. The ion center-to-center distance is therefore greater in this case and need not necessarily be limited to one interlinking solvent molecule.

The importance of ionic atmosphere versus ion association in a salt solution is largely dependent on the dielectric constant, ϵ, of the solvent. The onset of ion pair formation, where the mutual electrostatic stabilization of oppositely charged ions exceeds the energy of thermal collisions, occurs at lower ion concentrations in solvents of low dielectric constant. In polyethers, $\epsilon \sim 5$ to $10^{11,12}$; therefore extensive ion–ion interactions are expected to be favorable, particularly above 1 mol dm^{-3} (an O:Li ratio of 20:1 and less), which is the range of salt concentrations very often used for polymer electrolyte studies.

To consider models for the transport of ionic species in amorphous polymer electrolytes, the solution molarity should be considered. In systems of very low salt content, for example, <0.01 mol dm^{-3}, the conductivity should depend almost exclusively on the number of charge carriers. In this region, these are likely to be predominately free ions and the ionic mobility will be largely concentration independent. At the opposite extreme, in the ultraconcentrated medium, where the mean ionic separation is perhaps less than 0.5 nm, strong ion–ion interactions must exist and the polymer electrolyte may be best thought of as a "Coulomb fluid" having more in common with a molten salt or a molten hydrate than a nonaqueous solution. In a simple molten salt, the electric field around a quasispherical ion is centrosymmetric. Ion pairs thus cannot exist as stable entities because of the symmetrical electrostatic field surrounding the ions. Replacing a small proportion of the ions with solvent molecules destroys the symmetry and hence ion pairs may be stabilized. Between these two extremes, up to a concentration of approximately 1 mol dm^{-3}, widespread ion aggregation seems probable, possibly as ion pairs but with higher aggregates also of increasing importance as the salt concentration increases.[13,14] Bhattacharja[8] estimated a value of 82% ions in polyethylene glycol (PEG) solutions, $P(EG)_6LiCF_3SO_3$, to be ion paired at 165°C, and MacCallum et al.[13] calculated an approximate 77% pairs and 21% triples in low-molecular-weight $PEO–LiCF_3SO_3$ at salt concentrations of approximately 0.04 mol dm^{-3} at room temperature. So far as the conductivity mechanism is concerned, it is not certain whether center-of-gravity motion of the aggregates themselves or a transfer of ions between species occurs.

A statistical thermodynamic model has been developed to analyze the effects of ion pairing on the conductivity of polymer electrolytes.[15] In this model, the energy favoring complex formation (resulting from solvation and polarization) competes with an entropic term favoring the separate polymer plus contact ion pair. It was shown that the number of polymer-coordinated ions should decrease with rising temperature because of a favoring of entropic terms. This is in agreement with experimental findings where ionic aggregation[14] and eventual phase separation of the salt have been shown to occur.[16]

Cheradame and co-workers[17,18] analyzed conductivity data for network polymers by considering the electrolyte species as predominately associated. Quartets and ion pairs were assumed to coexist, and continuous ion pair transfer through the matrix was envisaged as resulting from exchange conformations assisted by segmental motion of the polymer as shown in Figure 9.1. It was assumed that the conductivity arises from the motion of free cations and anions, and therefore association and dissociation of aggregates are required to generate these charge carriers. A series of equilibria of the form

$$(MX)_2 \rightleftharpoons (MX)M^+ + X$$
$$(MX)_2 \rightleftharpoons 2(MX)$$
$$(MX)M^+ \rightleftharpoons (MX) + M^+ \tag{9.1}$$

shows how both positive and negative charge carriers may be generated. Higher aggregates were also considered. These equilibria were used to explain the rela-

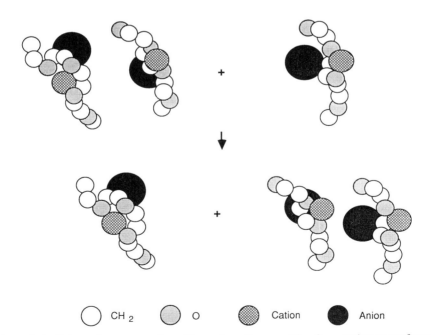

Figure 9.1. Schematic representation of ion pair motion, resulting from exchange conformations, assisted by segmental motion of the polymer.

tionship between charge carrier concentration and conductivity and variations in the transference number with concentration and temperature. In the aforementioned scheme, the positive triple ions formed are likely to be relatively immobile because more persistent ether oxygen–M^+ interactions would be expected; however, dissociation of aggregates that will form negative triple ion species cannot be discounted, and in this situation, the negative triple ion must be included as a charge carrier.

The interpretation of conductance–concentration dependence can be greatly simplified and hence information on ion–ion interactions can be derived by studying polymer electrolyte systems at very low salt content. Most of the available data relating to this concentration region have been obtained for low-molecular-weight polyether solutions.[13,19,20] Such systems, although similar, are unlikely to be fully representative of true solid polymer electrolytes because chain length differences may give rise to variations in ion–polymer and ion–ion interactions. More importantly, macroscopic displacements of the polymer chain can occur in liquid systems. The difficulty of carrying out such measurements in poly(ethylene oxide) (PEO) lies in the heterogeneous nature of the electrolyte and the residual catalytic impurities that give rise to a high background conductivity (see Chapter 2). A study has been carried out on a carefully purified amorphous polyether $[-(OCH_2CH_2)_x-OCH_2-]_y$ of average molecular weight 100,000.[21,22] Figure 9.2 shows the equivalent conductivity, Λ (ionic conductivity normalized by the total salt content), for this polymer containing $LiClO_4$ in the concentration range 8×10^{-4} to 1.2 mol dm^{-3} (up to an O:Li ratio of 15:1). This behavior is similar in form to that found for

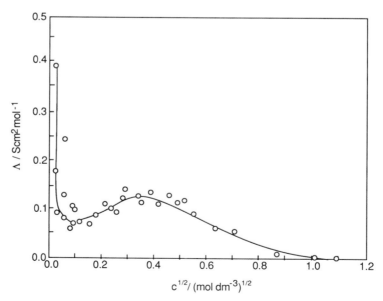

Figure 9.2. Concentration dependence of the equivalent conductivity for amorphous high-molecular-weight oxymethylene-linked PEO containing $LiClO_4$ at 25°C.

low-molecular-weight polyether systems[13,19–21] As the total salt concentration rises to about 0.01 mol dm^{-3}, a rapid fall in Λ is observed. Estimates for the reduction in ion mobility caused by ionic atmosphere effects show that this would be a very minor contribution to the fall in conductivity in this range. In addition, the T_g of these electrolytes up to approximately 0.1 mol dm^{-3} does not vary from the value for the pure polymer[22]; therefore, reduction in mobility through polymer chain segmental restrictions can also be discounted. It may be suggested that strong cation–anion interactions resulting in the formation of electrically neutral ion pairs occurs in this very low concentration region. Between 0.01 and approximately 0.1 mol dm^{-3}, the molar conductivity is seen to rise again and this may be explained by the formation of mobile charged clusters such as triple ions, by a progressive dissociation of ion pairs, or by a combination of both. Cameron et al.[23,24] suggested that ion pairs become progressively less stable with respect to free ions as the salt concentration increases, as the free ions are stabilized by an increasing ionic atmosphere. Such a situation may be expected in the ultraconcentrated range described above but is difficult to accept as being important at concentrations as low as 0.01 mol dm^{-3}. These data give substantial evidence for ion association as a predominant feature in solid polymer electrolytes at extremely low concentrations.

9.2. Spectroscopic Studies

Infrared (IR) and Raman spectroscopy have been used to study both interactions between the ions and host polymer and those between cations and anions. Raman lines cannot be observed from a pair of ions held solely by electrostatic forces but ion association can be determined from observations of the internal vibration modes of molecular ions. Variation in vibrational frequency and the Raman spectral linewidth can provide information about the influence of the environment on the molecular ions. Changes in dipole moment arising from an ion pair vibrations are Raman inactive but strong infrared absorptions, usually in the far-infrared are expected. These spectral features have been studied for polymer electrolytes.

An intense Raman-active band in the frequency range 860 to 870 cm^{-1} is observed for many PEO–salt and crown ether complexes[25,26] but is not observed for the pure polymer. This absorption has been assigned to a metal–oxygen breathing motion and because it is not IR active, it can be implied that the cation is symmetrically surrounded by ether oxygens. This band appears to be IR active when a mixture of PEO and acetonitrile is used as solvent.[27] Studies of LiAsF$_6$ and LiClO$_4$ have been carried out in this solvent using IR and ultrasonic relaxation techniques and indicate that at salt concentrations above approximately 0.3 mol dm^{-3}, specific anion effects are apparent that may reflect competition with the solvent for cation coordination. Variations in this IR band with molecular weight were explained by different polymer–ion conformations, dependent on the ability of the polymer to wrap around the cation.

Far-infrared (100–400 cm^{-1}) studies of a number of P(EO)$_4$–Na salt complexes show what has been termed "solvent cage"-type vibrations, corresponding to the motion of the cation relative to the anion and surrounding ether oxygens. It can be attributed to a combination of O—C—C, C—O—C bending and C—C torsional modes.[14] The spectra of NaBr, NaI, NaSCN, and $NaBF_4$[25] complexes are very similar in this region but $NaCF_3SO_3$ gives rise to completely different bands and significant changes are observed when BD_4^- is substituted for BH_4^-. This anion dependence has been regarded as arising from ion pair interactions. Although Papke et al.[28] reported the IR and Raman absorptions for $LiCF_3SO_3$ and $LiBF_4$, this frequency range was not covered and comparisons with Na complexes could not be made. Narrowing of the band width occurs with increased conformational order and decreases in chain flexibility. These also give rise to increases in peak frequency.[29] Roman spectra of the complexed polymer are thought to be comprised of two super-imposed bands; one broad, and arising from uncomplexed polymer and an additional mode with a much smaller band width. As the temperature is raised, this latter feature becomes less prominent, because ion—ether oxygen interactions decrease (Figure 9.3). This is consistent with salt precipitation from solution encountered at high temperatures as the temperature and concentration of salt are increased on PEO and PPO. This is more noticeable in PPO which forms weaker cation–polymer interactions.[16] Papke et al.[28] report minimal changes in the band features for PEO–Na

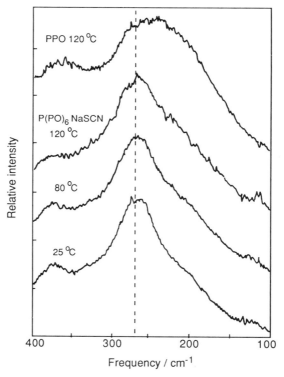

Figure 9.3. Raman scattering spectra of PPO–NaSCN and PPO at various temperatures. From Ref. 29.

salt complexes as temperature is raised, which would be expected for these more stable complexes.

Spectroscopic techniques were used by Papke and co-workers[30,31] to study the influence of ion pairing in PEO–NaBF$_4$ and PEO–NaBH$_4$ electrolytes. The high-frequency B—H stretching vibration at 2250 cm^{-1} is readily perturbed by changes in the environment or geometry and is thus a useful probe of the anion local environment. The vibrational band structures of the BH$_4^-$ ion in the polymer were complex and were interpreted as an indication of the anion symmetry being lowered from a tetragonal arrangement. The analogous BF$_4^-$ Raman bands were essentially the same as those for aqueous or molten NaBF$_4$ which are known to arise from tetragonal symmetry. The conductivity of the NaBH$_4$ electrolyte is considerably lower than that of the equivalent NaBF$_4$ system and it was suggested that these findings implied strong ion pairing in the former material, while the latter appears to exist as either solvated ions or as a lattice structure like the molten salt rather than as ion pairs (or an immobile strongly interacting environment) in PEO. Results of a

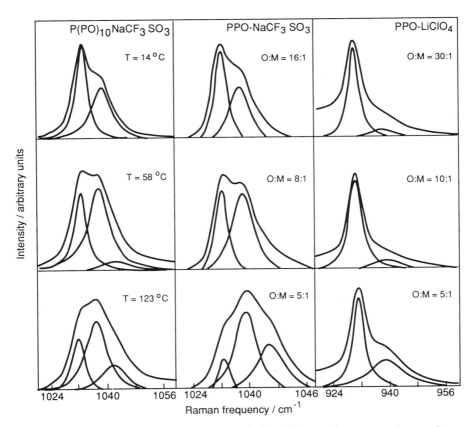

Figure 9.4. Anion symmetric stretching mode for different salt concentrations and temperatures of PPO–NaCF$_3$SO$_3$ and PPO–LiClO$_4$. From Ref. 42.

Raman spectroscopy study of a liquid P(PO)$_7$NaSCN electrolyte were also consistent with strong ion–ion interactions.[29] The frequencies of the CN (2063 cm^{-1}) and SC (755 cm^{-1}) stretching vibrations differed from values found for fully solvated ions in THF or DMF (2053 and 735 cm^{-1}) but were close to the values 2066 and 754 cm^{-1} found for ion pairs in these solvents.[32,33] Furthermore, cationic substitution altered these frequencies which would not be expected if SCN$^-$ ions were not ion-paired in PPO solution.

Information on ion interactions has been obtained from observations of symmetric stretching modes of the internal vibrations of CF$_3$SO$_3^-$ anion in PPO–NaCF$_3$SO$_3$[34,35] and PEO–NaCF$_3$SO$_3$[36] and ClO$_4^-$ in PPO.[35,37] Figure 9.4 shows the SO$_3$ and ClO$_4$ symmetric stretching bands. The band at 1032 cm^{-1} was assigned to "free" trifluoromethanesulfonate ions, ion-paired trifluoromethanesulfonate at 1037 cm^{-1}, and clusters at approximately 1046 cm^{-1} by comparison with bands arising in NaCF$_3$SO$_3$–CH$_3$CN solutions. The Raman spectra of PEO– and PPO–NaCF$_3$SO$_3$ complexes show approximately similar behavior but with an indication of slightly higher "free" ion concentration in PPO which is contrary to what may be predicted.[36,38] PPO–LiClO$_4$ complexes show quite different behavior in that only two absorptions are found in the frequency region for the anion. The intensities are again concentration and temperature dependent which may indicate the formation of precursors for salt precipitation.

Frech et al.[39] reported the vibrational spectra of the thiocyanate ion in PPO–NaSCN, PPO–KSCN, and PPO–LiSCN solution. Bands arising from the thiocyanate internal vibrations are observed as single symmetric features for the former two which were assigned to ion-paired thiocyanate; the latter showed asymmetry, implying a more complex interaction.

9.2.1. General Interpretation of Spectral Data

It is possible to interpret the preceding data in a more general manner.[40] The concentration of salt in the polymer is of the order of 0.6 to 4 mol dm^{-3} (30:1–5:1), and it is more realistic to view these systems as "Coulomb fluids," having much in common with molten salt hydrates. Table 9.1 shows Raman spectral information for various environments of nitrate, perchlorate, and trifluoromethanesulfonate ions. In comparing data for polymer electrolytes with those for aqueous solutions of these salts it should be remembered that the dielectric constant of the solvent is approximately 5 in the former case as compared with approximately 80 for water and thus much more extensive ion association will be expected for the polymeric systems; however, spectral variations arising from near-neighbor influence may be expected to follow similar trends.

The Raman spectra for aqueous solutions of LiNO$_3$ have been well studied at all concentrations.[41] As the salt concentration increases, some absorptions are seen to be sensitive to gradual change; others show the coexistence of two environments above a particular salt content. For example, from Table 9.1, the symmetric stretching vibration systematically decreases from 1067 cm^{-1} for the melt to a dilute

Table 9.1. Raman Frequencies for Molecular Anions in Various Phases or Solutions

Absorption Frequency (cm^{-2})

Salt	Phase/Solvent	Concentration mol dm⁻³	Concentration O:M	Ref.	$\nu_1(NO_3)$	$\nu_3(NO_3)$	$\nu_4(NO_3)$
$LiNO_3$	Crystalline	—	—	41	1071	1384	735
	Melt	—	—	41	1067	1467, 1381	737, 714
	H_2O	9	—	41	1055	1458, 1389	740, 720
	H_2O	<5	—	41	1049	1429, 1352	720
	PEO	6	4:5:1	28	1045	1416, 1324	729, 719

Salt	Phase/Solvent	mol dm⁻³	O:M	Ref.	$\nu_1(ClO_4)$	$\nu_3(ClO_4)$	$\nu_4(ClO_4)$
$NaClO_4$	Crystalline	—	—	45	939	1081, 1111, 1124	625, 634
	H_2O	Dilute	—	45	935	1102	628
$LiClO_4$	PPO	0.9	30:1	37	932, 940	—	—
	PPO	2.7	10:1	37	932, 940	—	—
	PPO	5.4	5:1	37	932, 940	—	—

Salt	Phase/Solvent	mol dm⁻³	O:M	Ref.	$1A_1(\nu_{c-s})$	$E(\rho, SO_3)$	$E(\delta_d CF_3)$	$A_1(\delta_s SO_3)$	$A_1(\delta_s CF_3)$	$A_1(\nu_s SO_3)$	$A_1(\nu_s CF_3)$
$NaCF_3SO_3$	H_2O	4	—	46	321	353	520	580	766	1038	1230
	CH_3CN	0.1	—	34	313, 316	—	—	—	755, 758	1034, 1039	—
	CH_3CN	1.0	—	34	313, 316	349, 354	—	—	755, 758	1034, 1039, 1045	—
	CH_3CN	1.5	—	34	316	349, 352	—	—	758, 760	1035, 1040, 1045	—
	PPO	0.9	30:1	34	313	348, 352	517	575	756	1032, 1037, 1042	1226
	PPO	2.7	10:1	34	314	348, 352	518	575	756	1032, 1038, 1042	1227
	PPO	5.4	5:1	34	316	348, 352	519	576	759	1033, 1039, 1046	1229

solution value of 1049 cm^{-1}. In concentrated aqueous solution, four bands are found at 1352, 1429, 1389, and 1458 cm^{-1}, the former two characteristic of dilute systems, the latter two similar in frequency to those found in the molten salt. The absorption found at 720 cm^{-1} in dilute solutions splits at concentrations above approximately 9 mol dm^{-3} to give a second band at approximately 740 cm^{-1} which increases in intensity with concentration. At this concentration, packing considerations require anion–cation contact. This splitting is also observed in $P(EO)_{4.5}LiNO_3$ systems.[28] Difficulty exists in further assigning bands for the polymeric system because of the overlap of ion- and polymer-based absorptions. Clancy et al.[41] studied the interaction of $AgNO_3$ in poly(pentamethyl sulfide) by IR spectroscopy. The nitrate ion band at 1050 cm^{-1} is only Raman active but can be seen in the solid-state IR spectrum of some nitrate salts, where its appearance is attributed to the distortion of the ion from D_{3h} symmetry and a shift or splitting of the band is indicative of ionic interaction. Splitting of this band was observed in highly concentrated polymer–salt solutions (6:1 to 1:1 S:Ag$^+$). Space-filling models show, however, that the polymer cannot form favorable conformations to allow cation coordination to two or more adjacent sulfur atoms on the polymer chains.

Spectral features of polyether solutions of $MClO_4$ (M = Li, Na) and $NaCF_3SO_3$ in various solvents are compared in Table 9.1. The higher of the two frequency lines arising from the symmetric stretching mode of the ClO_4^- ion in PPO (940 cm^{-1}) is very close in value to that for crystalline perchlorate salts. As this absorption line is known to increase in intensity with temperature[14,35,43] (and salt precipitation is the ultimate consequence of raising the temperature[16,29]) it may be preferable to consider it as arising as a result of the anion experienceing a lattice-type environment rather than as a result of discrete ion pair formation. The effects of polymer and ion interactions are again seen in the far-infrared where the absorption at approximately 234 cm^{-1} due to poly(propylene glycol (PPG), shifts to approximately 266 cm^{-1} and the width decreases sharply on adding salt to the polymer. By raising the temperature the absorption band shifts back toward that for pure polymer. Thus as O . . . M$^+$ interactions decrease, it can be assumed that M$^+$ . . . X$^-$ interactions increase, increasing the likelihood of the anion experiencing a more lattice-like environment.

For PPO–$LiClO_4$ systems, a second more intense line arising from the symmetric stretching mode is observed and has a frequency value close to that for perchlorate in dilute aqueous solution (935 cm^{-1}). Although this may be interpreted as resulting from free ions in polymer solution, Hester and Plane[44] and Ross[45] have shown for numerous sulfates and perchlorates that even in highly concentrated media, the Raman spectrum remains insensitive to the environment *even though other methods have shown the existence of ion association.* The effects on this Raman line in lower-dielectric-constant solvents such as acetonitrile are uncertain, but it is probable that a single absorption would again be found, at least in dilute solution. This absorption in PPG solution therefore cannot be assigned to a particular environment with any certainty.

Data for the trifluoromethanesulfonate ion in various media are shown in Table

9.1. The symmetric stretching band is split into three lines in PPO solution in this case. If the highest-frequency absorption is taken at 1046 cm^{-1}, it can be assigned, as for the highest-frequency absorption for ClO_4^-, to a latticelike vibration of the molecular ion. The absorption data in Table 9.1 for a 4 mol dm^{-3} aqueous $NaCF_3SO_3$ solution (1038 cm^{-1}) cannot therefore represent the Raman absorptions of a lattice structure; however, Miles et al.[46] suggest that cation–trifluoro-methanesulfonate ion coordination is not significant in aqueous solution. In acetonitrile solution, which has a lower dielectric constant than water, ionic interactions are more probable. The SO_3 symmetric stretch absorption in this medium (as are the other absorptions) is split into three lines, one that corresponds in this instance to the 1046 cm^{-1} line found in PPG solution. Ion—ion interactions, if strong, may cause splitting of degenerate vibrational modes by lowering the symmetry of the molecular species. Trifluoromethanesulfonate is, however, an unusual anion in that it is possible that there is marked delocalization of π electron density in the anion which involves the lone-pair oxygen π orbitals, the vacant sulfur d orbitals, and possibly the vacant antibonding CF_3 system. Anion–anion interactions may therefore account for some of the Raman features. On the other hand, ^{19}F NMR studies[47] suggest that the CF_3 group does not interact to any extent with the surrounding species.

^{23}Na and ^{19}F NMR measurements of a $P(PO)_{30}NaCF_3SO_3$ substantiate some of the information obtained by Raman spectra.[47] Measurements of the ^{23}Na nuclei reveal a two-component band: a relatively narrow line (1 kHz at 273 K) superimposed on a broad line ($\sim$10 kHz). The spin–lattice relaxation time for the narrow line is very short ($T_1 \sim 9$ μs at 297.5 K), and the broad band has a value of $T_1 \sim 13$ ms at the same temperature. The ^{19}F line has a relatively small linewidth (350 Hz at 273 K) and only one component which narrows dramatically with increasing temperature. Greenbaum et al.[48] investigated the system $P(PO)_8NaCF_3SO_3$ using ^{23}Na NMR and found essentially the same. These results suggest that the sodium ions exist in two environments with differing dynamic properties.

9.3. Transference Numbers

Although transport properties have received a good deal of study, no rigorous interpretation of data from various transport experiments has been carried out. In the majority of studies, it has been assumed that the only mobile species are cations and anions, M^+ and X^-, and experimental data have been interpreted in such a way as to give transport numbers, t_i. From the preceding discussion, there is growing evidence that ion association is prevalent over an extremely wide concentration range. Whether these species are mobile or not cannot be established from conductivity data but in many instances it may not be valid to quote transport numbers.

Transference numbers T_i are defined[49] as the net number of Faradays carried by ion constituent i (rather than ion i) in the direction of the cathode or anode on the passage of 1 Faraday of charge across the cell. T_i and t_i are only equal when the electrolyte dissociates into two ionic species. For an associated system, containing only M^+, X^-, MX, M_2X^+, and MX_2^-, the transference number of the X constituent, T_X, may be related to the individual transport numbers by

$$T_X = (t_{X^-} + 2t_{MX_2^-} - t_{M_2X^+})$$

A similar equation for the cation transference number, T_M, can be given and $T_M + T_X = 1$. On the assumption that polymer electrolytes in which both cations and anions are free to move may contain mobile associated species, it is more appropriate to refer to transference numbers when data analysis does not assume a fully dissociated electrolyte.

A variety of techniques have been employed to probe transport processes in polymer electrolytes, but so far there are considerable variation and disagreement concerning the values of transference numbers. For example, for PEO–LiClO$_4$ electrolytes at a salt content of approximately 8:1 to 12:1 and at temperatures around 120°C, a number of techniques point to a Li$^+$ transport number in the range 0.2 to 0.3.[50–55] The effects of temperature and concentration on the magnitude of the transport number are not obvious. For PEO network–LiCF$_3$SO$_3$, Tubandt measurements[49–51] at 24:1 and 70 to 120°C gave a cationic transference number of approximately 0.2, independent of temperature. Direct-current polarization techniques[56,57] and concentration cell techniques[55] gave higher values while and pulsed field gradient measurements indicated a highly temperature-dependent number of approximately 0.2 for P(EO)$_8$LiCF$_3$SO$_3$ at 120°C. Other lithium salts in PEO yield cationic transference numbers in the range 0.17 to 0.58.[55–58] These conflicting results may be attributed in part to the use of a model of a fully dissociated polymer electrolyte which is often used to interpret measurements. It is also important to realize that unlike the Tubandt/Hittorf and concentration cell methods where only charged species are involved in the measurements, dc polarization and pulsed field gradient techniques are affected by neutral species and therefore do not give t_i or T_i if *mobile* ion pairs or higher neutral aggregates are present.

Bruce and Vincent[59,60] have examined the effects of ion association on the methods used to study ion transport, showing that the various techniques are influenced in different ways by different species. Illustration was made for all-amorphous polymer electrolytes which contain a univalent salt MX and the species present in the electrolyte were restricted to M^+, X^-, MX, M_2X^+, and MX_2^-. This is discussed later.

9.3.1. Transport Numbers from Diffusion Coefficient Determinations

Transport numbers may be evaluated through ionic mobility and by non-electrochemical techniques if ionic diffusion coefficients can be determined. Diffusion techniques involve measurement of the flux of both charged and electrically neutral species. Serial sectioning radiotracer and pulsed field gradient NMR techniques have commonly been applied to polymer electrolytes to obtain diffusion coefficients and, subsequently, the ionic transference numbers. As discussed later, these techniques are valid only if the electrolyte is fully dissociated. Chadwick and Worboys have described these applications.[61]

Bruce and Vincent[59] illustrated the effects of ion association on diffusion studies by considering the radiotracer technique, but it is applicable to all diffusion measurements. The total flux of a radioactive species $*M$ into the material from a thinly deposited surface layer may be given as

$$J_{*M} = -D'_{*M}\frac{d*c}{dx}$$

$$= (-\alpha_1 D_{*M^+} - \alpha_3 D_{*MX} - \alpha_4 D_{*M(M)X^+} - \alpha_5 D_{*MX_2^-})\frac{d*c}{dx}$$

where $*c$ is the total radioactive salt concentration and the concentration of each species may be expressed as

$$[*M^+] = \alpha_1*c, \ [*MX] = \alpha_3*c, \ [*M(M)X^+] = \alpha_4*c, \ [*MX_2^-] = \alpha_5*c$$

D'_{*M} is the diffusion coefficient obtained experimentally for $*M$. A similar expression for an anionic radiotracer flux may be established:

$$J_{*X} = -D'_{*X}\frac{d*c}{dx}$$

$$= (-\alpha_2 D_{*X^-} - \alpha_3 D_{M*X} - \alpha_4 D_{M_2*X^+} - \alpha_5 D_{M*X(X)^-})\frac{d*c}{dx}$$

When the concentration or mobility of associated species is negligibly small compared with that of the free ions then D'_{*M} and D'_{*X} correspond to the diffusion coefficients of the respective free ions and, on application of the Nernst–Einstein relationship, transport numbers may be defined as

$$t_+ = \frac{D_+}{(D_+ + D_-)} \qquad t_- = \frac{D_-}{(D_+ + D_-)}$$

If, however, the sum of the terms for associated species is large in comparison with the terms for the free ions, then $J*M \sim J*X$ and $D'_{*M} \sim D'_{*X}$. Using the above equations, t_+ and t_- are both approximately 0.5.

9.3.1.1. Radiotracer Studies

Chadwick and co-workers[62–65] used this technique to study the diffusion of ^{22}Na, ^{86}Rb, and $S^{14}CN$ in PEO– and PPO–salt systems at O:M ratios of less than 20:1. The transport mechanism was shown to be complex but it was clear that the cationic and anionic diffusion coefficients were comparable, as shown in Figure 9.5, which may point to high mobility and/or concentration of associated species.

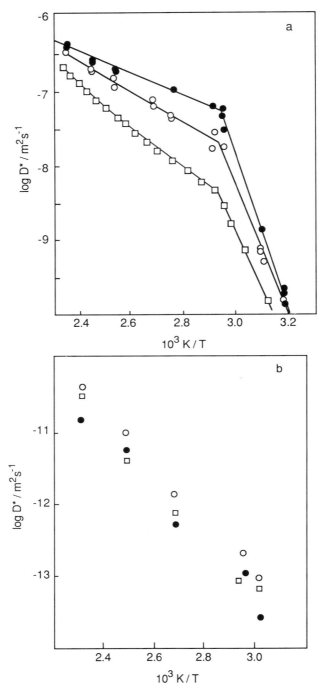

Figure 9.5. Diffusion coefficients as a function of temperature, determined through radio-tracer experiments. (**a**) P(EO)$_8$NaSCN: (□) calculated values from conductivity measurements prior to correction for crystalline phases, (●) S^{14}CN, (○) ^{22}Na. From Ref. 61. (**b**) P(PO)$_8$NaSCN: (●) ^{22}Na, (□) S^{14}CN, (○) $D^*_{(total)}$. From Ref. 63. Reproduced by permission of the SCI.

9.3.1.2. Pulsed Field Gradient NMR

As in radiotracer experiments, pulsed field gradient NMR (pfg-NMR) gives a direct measurement of the diffusion coefficient, although it is limited to relatively fast diffusion. The above arguments hold here and interpretation of data is made more difficult if associated, mobile species are involved. Measurements in PEO and PPO systems containing $LiCF_3SO_3$ and other lithium perfluorosulfonates or $LiClO_4$ have been carried out by following the motion of labeled ^7Li and ^{19}F ions.[66–70] Diffusion coefficient measurements for $P(EO)_8LiCF_3SO_3$ and $P(EO)_8LiClO_4$ are summarized in Figure 9.6. Unfortunately, direct comparison of the data is not possible as different polymer molecular weights (4×10^6 and 6×10^5) and temperature ranges were studied, although the former should have a negligible effect. The extrapolated values of Bhattacharja et al.[67] for both D_{Li} and $D_{CF_3SO_3}$ lie considerably below other reported results,[70] although the conductivities were comparable. Lindsey et al.[68] report a cationic transference number of 0.4 to 0.5 which was almost invariant in the temperature (80–170°C) and concentration (20:1–6:1) ranges of study. Bhattacharja et al. gave values of 0.3 to 0.4 over the temperature range 140 to 165°C for an 8:1 sample. Gorecki et al.[71] examined Li diffusion in PEO–$LiClO_4$ and PEO–$LiC_{p+1}F_{2p+3}SO_3$ ($p = 5,7$) over the concentration range 20:1 to 6:1. Pulsed field gradient NMR measurements were combined with conductivity data to estimate the Li transport number. The former study gave values of approximately 0.2 to 0.3, and the latter produced a cationic transport number of approximately 0.3 at 120°C and approximately 0.45 at 140°C. Some consistency has been found between measured

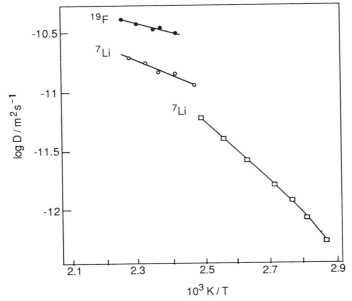

Figure 9.6. Pulsed field gradient NMR data for $P(EO)_8LiCF_3SO_3$ ($\bullet,\bigcirc$)[67] and $P(EO)_8$ $LiClO_4$ ($\square$).[69]

conductivity values and those deduced from pfg-NMR and the Nernst–Einstein equation. The ratio $\sigma(T)/D(T)$ was reported to be constant for 6:1 and 8:1 PEO–LiCF$_3$SO$_3$ systems, but deviation occurred at a concentration ratio of 20:1.[54] The constancy of this ratio may indicate that neutral species are in fact immobile in polymer electrolytes,[72] unlike the situation proposed earlier; however, the transport numbers in polymer electrolytes, as measured through diffusion techniques, tend toward a value of approximately 0.5, as predicted for systems containing mobile ion pairs. Insufficient studies of conductivity and ion diffusion have been reported to draw any firm conclusions.

Estimates of the diffusion coefficients of Li- and Na-complexed salts from relaxation time measurements have been carried out by a number of groups.[73-77] It is not straightforward to relate linewidth to diffusion coefficients, primarily because it involves trying to relate NMR parameters to long-range diffusion as distinct from short-range motional disorder. Further problems are caused by quadrupolar effects and the fact that most studies were carried out in heterogeneous systems.

9.3.1.3. Electrochemical Determination of Diffusion Coefficients

Diffusion coefficients and ionic mobilities may be determined electrochemically by chronoamperometry and chronopotentiometry. Such measurements were made on PPG–NaCF$_3$SO$_3$ (40:1) doped with millimolar quantities of electroactive Ag$^+$ and I$^-$.[78] The diffusion coefficients of Ag$^+$ and I$^-$ were found to be approximately equal; however, interactions between Ag$^+$ and I$^-$ and the supporting electrolyte are likely, forming clusters such as AgNaCF$_3$SO$_3{}^+$ and NaCF$_3$SO$_3$I$^-$ and therefore a change of supporting electrolyte could markedly influence the diffusion coefficient.

Sørensen and Jacobsen[79] measured cationic transference numbers of P(EO)$_x$ LiCF$_3$SO$_3$ ($x = 2.5-16$) through the observation of limiting currents on chronoamperometric experiments. Cationic transference numbers were found to be <0.1 below 85°C, increasing to 0.35 to 0.4 at 125°C. At lower temperatures, <120°C, a large percentage of material is present as crystalline phase, as can be seen in the phase diagram in Chapter 4, and this will undoubtedly affect diffusion measurements. Similar measurements were carried out by Watanabe et al.[80-82] for PEO–LiClO$_4$, PPO–LiClO$_4$ networks, and poly(ethylene succinate)–LiSCN. A cationic transference number of approximately 0.5 was estimated for PEO–LiClO$_4$, and the thiocyanate-containing material produced an estimated value of 0.91 to 0.99. Ionic mobilities and diffusion data were derived assuming complete dissociation of the electrolyte. In addition, it may be anticipated for these methods that reactions at the electrodes, for example, formation of passivating surface layers or adsorption, could affect the transients and therefore interpretation purely in terms of bulk properties may not be valid. Watanabe et al.[83] have determined diffusion rates of ferrocene derivatives dissolved in an amorphous crosslinked PEO-based polymer complexed with LiClO$_4$. They were measured by a technique that detects the rate of transport to an oxidizing microdisc electrode. The diffusion coefficients 3×10^{-7} to 2×10^{-8} cm^2 s^{-1}, depending on the size of the ferrocene derivative, decreased with increasing ferrocene concentration, increasing LiClO$_4$ concentration, and decreasing temperature. The qualitative similarities between temperature and concentration dependencies for diffusion rates and ionic conductivity behavior suggest that high electrolyte concentration, leading to decrease in ionic conductivity, can be related to a decrease in ionic diffusivity. Wrighton and co-workers[84] reported the use of microelectrode arrays for direct measurement of diffusion of Ag^+ dissolved as $AgCF_3SO_3$ in $PEO-LiCF_3SO_3$ and $MEEP-LiCF_3SO_3$. The experiment involves anodically stripping Ag from an Ag-coated Pt microelectrode and electrochemically detecting Ag^+ at nearby Pt electrodes on its reduction. The time dependence of the collector current for $Ag^+ \rightarrow Ag$ after generation of Ag^+ allows evaluation of the diffusion coefficient. D_{Ag} values of 5×10^{-9} cm^2 s^{-1} (at 25°C) and 7×10^{-8} cm^2 s^{-1} (at 79°C) were measured for MEEP and PEO systems, respectively. The nature of the diffusing species could not be established.

9.3.2. Measurement of the Transport of Charged Species Only

9.3.2.1. Hittorf/Tubandt Method

This technique is based on Faraday's laws and requires the determination of composition or weight variations in the electrolyte regions near the two electrodes, caused by the passage of a measured quantity of charge through the cell. For a cell of the type

$$M_{(s)} | \text{polymer-}MX_{(s)} | \text{polymer-}MX_{(s)} | \text{polymer-}MX_{(s)} | \text{polymer-}MX_{(s)} | M_{(s)}$$

where the polymer compartments are identical, 1 mol of M is stripped from the anode and deposited on the cathode on passing 1 Faraday of charge. Current is carried by the motion of M^+, M_2X^-, X^-, and MX_2^-. Ion pairs are not involved in the flux provided the central compartments remain invariant throughout the experi-

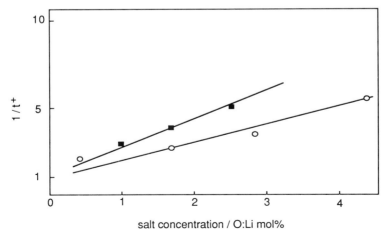

Figure 9.7. Plot of the reciprocal transport number versus salt concentration for two different PEO networks, PEO(1000) (○) and PEO(2000) (■), at 90°C. From Ref. 17.

ment. For an associated electrolyte, analysis of the cathode compartment leads only to the determination of the net transfer of X due to the transport of M_2X^+ in and X^- and MX_2^- out of the compartment. It is very difficult to apply this technique to polymer electrolytes because of the necessity to maintain the electrolyte as non-adherent thin film compartments. Cheradame and co-workers[51] have applied this technique to highly crosslinked networks. The transport number of the cation was shown to be unity when the anion was immobilized.[1] A cation transference number of 0.2 was established for a PEO(1000)–LiClO$_4$-based membrane[51] for O:Li = 20:1. An increase in salt concentration decreases the cation transference number, which is also sensitive to the PEO molecular weight between chain crosslinks, decreasing as the chain lengths increase. By assuming a model based on the equilibria shown in Eq. (9.1) a linear relationship between reciprocal cation transference number and salt content was predicted.[17] This is shown in Figure 9.7 to hold reasonably well for PEO networks. PPO–NaBPh$_4$ complexes show a reverse effect that is thought to be the result of variations in interactions between the ions and PEO, PPO, and urethane linkages.[85]

9.3.2.2. Concentration Cell Techniques

For a cell of the form

$$M_{(s)} \ | \ \text{polymer–}MX_{(s)} \ \| \ \text{polymer–}MX_{(s)} \ | \ M_{(s)}$$

$$c_1 \qquad\qquad\qquad\qquad c_2$$

for $c_1 > c_2$, the ionic species are transferred across the boundary in amounts determined by their concentrations and transference numbers. The reversible discharge of the cell involving the passage of 1 Faraday would correspond to the transfer of T_M mol of M-containing species and T_X mol of X-containing species across the junction in opposite directions, the net result being the transfer of T_X mol of salt. For c_1 and c_2 to differ by an infinitesimal amount,

$$dE_{\text{cell}} = -(RT/F)T_X/d \ ln \ a$$

or

$$T_X = -(F/RT)dE_{\text{cell}}/d \ ln \ a$$

where a is the activity of the salt. It is necessary to know how the activity of the salt varies with concentration. Armand and co-workers measured transference numbers of LiI in PEO by this method.[55] Mean salt activities were determined by making electromotive force (emf) measurements of the cell

$$Li_{(s)}| \ PEO–LiI_{(s)} \ | \ PbI_{2(s)} \ | \ Pb_{(s)}$$

and a transference number T_{Li} was determined to be 0.34 ± 0.06 at 90°C. On the assumption that LiI and LiClO$_4$ have similar activities, T_{Li} in PEO–LiClO$_4$ electrolytes was calculated to be 0.25 ± 0.07. Pb|PbCF$_3$SO$_3$ was used as a trifluoromethanesulfonate reversible electrode to determine the mean activities for PEO–

$LiCF_3SO_3$. A value of $T_{Li} = 0.7 \pm 0.1$ was found which was surprisingly high. This was interpreted as being the result of the different nature of mobile species compared with the former examples. The same authors determined the cationic transference number in PEO–$LiClO_4$ by using poly(decaviologen) (DV),[86]

$$-[-(CH_2)_{10}- \overset{+}{N} -\hspace{-0.3em}\bigcirc\hspace{-0.3em}-\hspace{-0.3em}\bigcirc\hspace{-0.3em}- \overset{+}{N} -]-$$
$$X^- X^-$$

poly(decaviologen)

as an anion-specific electrode and measuring emf values for cells 1 and 2 with transference,

$$Li_{(s)}| \, P(EO)_8LiClO_{4(s)} \, \| \, P(EO)_{>8}LiClO_{4(s)}| \, Li_{(s)}$$

and without transference,

$$Li_{(s)}|P(EO)_8LiClO_{4(s)}|poly(DV^{2+},2ClO_4^-),poly(DV^+,ClO_4^-)|P(EO)_m|Li_{(s)}$$

By plotting the emf values of cell 1 against cell 2, $T_{ClO_4^-}$ may be determined from the slope of the resulting straight line. The value of T_{Li} was calculated to be 0.21 ± 0.08.

9.3.2.3. Cells in Force Fields

If a symmetric cell of the form

$$M_{(s)}| \, polymer-MX_{(s)} \, | \, M_{(s)} \tag{9.2}$$

is placed radially on a spinning rotor an emf develops because there is a difference in kinetic energy between the reagents at the two electrodes. MacInnes and Ray[87,88] developed the technique for aqueous solution electrolytes but it has not received wide application. T_X may be obtained by measuring the cell potential as a function of rotation speed. This method avoids the inherent problems of using a solid electrolyte experienced in the above two methods and is currently being studied in our laboratories.

9.3.3. Transport under a Chemical Potential and Electrical Gradient

9.3.3.1. Alternating-Current Impedance

Sørensen and Jacobsen[89] suggested a technique for measuring transport numbers based on the theory of MacDonald[90,91] which involves analysis of the ac impedance spectra of the symmetric cell in Eq. (9.2). For the idealized case, the equivalent circuit and response of Figure 9.8 may be anticipated. The high-frequency semicircle is due to cation migration through the bulk and dielectric polarizations. At lower frequencies, a semicircle arises as a result of the charging and discharging of the

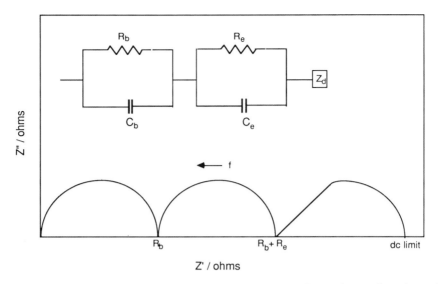

Figure 9.8. Idealized equivalent circuit and complex plane plot for a polymer electrolyte cell with electrodes that are nonblocking to the cation. R_e, electrode resistance; C_e, electrode capacitance; R_b, electrolyte resistance; C_b, bulk or geometric capacitance; Z_d, diffusion-controlled impedance.

electrode–electrolyte interface and reactions of the electroactive ion at the electrode interface. At lowest frequencies, the current is affected by concentration gradients that gives rise to diffusion in the electrolyte, and a skewed semicircle that approaches the limiting dc value of the cell impedance is observed.

For fully dissociated electrolytes, the cation transference number T_+ can be evaluated by comparison of the width of the skewed semicircle, Z_d, with the value of the bulk resistance, R_b:

$$T_+ = 1/(1 + Z_d/R_b)$$

The ac technique has been by far the most widely used method for obtaining transference numbers on polymer electrolytes,[89,92–95] but again the analysis is relevant only when mobile associated species are not present. In addition, electrode phenomena such as finite electrode kinetics or formation of passivating layers are not considered and, as suggested by Fauteux,[96] the diffusion responsible for the low-frequency arcs may not be a bulk process but one of the passivating film on the electrode surface which is generally observed when polymer electrolytes are in contact with metallic lithium (see Chapter 10).

Two mathematical models have been described[97,98] that can be applied to the transport of species through a thin layer of polymer electrolyte contained between nonblocking electrodes. Expressions have been derived to describe the ac impedance response that may arise when certain conditions relating to the components present in the system are considered. Lorimer[97] derived flux equations from nonequilibrium thermodynamics and used them to deduce appropriate differential equations for transport in thin-layer lithium cells. The theory was applied to two systems: (1) solvent containing a fully dissociated electrolyte, (2) solvent containing an electrolyte that is partially ionized. The situations of fast and slow ion pair equilibration compared with the rate of diffusion were each considered. Diffusional impedance equations were derived that included correction terms to the convertional MacDonald equation[90,91] for concentrations other than infinite dilution. The inclusion of ion pairs as mobile species in the electrolyte gives rise to additional terms in the diffusional impedance. This can give rise to marked distortion of the diffusion arc on impedance plots. The combined diffusion and chemical equilibrium arcs in Figure 9.9 may be distinct or combined to form an arc similar to a single diffusion arc. The effect of slow equilibration therefore was found to produce a slightly depressed and contracted finite Warburg diffusion impedance.

Pollard and Compte[98] considered the same electrolyte components but, in addition, evaluated the effects on an inert crystalline phase distributed through the

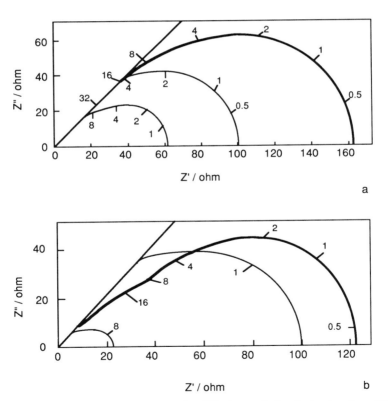

Figure 9.9. Calculated impedance plots for diffusion and chemical relaxation. Diffusion coefficient $D = 2 \times 10^{-9}$ m^2 s^{-1}, cell thickness $L = 2 \times 10^{-5}$ m. (—) Total impedance, (—) pure diffusion. The small semicircle is relaxation. (**a**) Relaxation rate constant $k' = 10$ s^{-1}, (**b**) $k' = 100$ s^{-1}. Pure diffusion arc normalized to 100 Ω. From Ref. 97.

electrolyte. A multicomponent transport equation was used that included the effects of diffusion, migration, and convection and interactions among the species fluxes. It was concluded that meaningful diffusion coefficients and transference numbers could be obtained only if precise information was available on the species present in the electrolyte and the nature and rates of interfacial reactions. With the model, consistency tests were established and were used to distinguish between dilute and concentrated electrolytes and, furthermore, to determine when it was inappropriate to treat the material as a binary electrolyte and to include higher-order species in the model. Impedance data for $P(EO)_{16}LiCF_3SO_3$ and $P(EO)_8LiClO_4$ were compared with those predicted from a simple binary electrolyte model and good agreement was observed; however, the consistency checks implied that ion pairing and/or higher aggregate formation should be expected in these electrolytes which raises the question of the sensitivity of impedance data to a more appropriate transport model.

9.3.3.2. Direct-Current Polarization Methods

Fully Dissociated Electrolytes. The basis of this method is exactly the same as that for the technique of Sørensen and Jacobsen,[89] described in the previous section, who applied an ac signal to this type of symmetric cell. At very low frequencies, the cell impedance approaches the limiting dc value. Bruce and Vincent[99] considered the symmetric cell in Eq. 9.2 which is polarized by the application of a constant potential difference between the electrodes. It was assumed that the electrolyte was fully dissociated, convection did not contribute to ion transport, and charge transfer processes at the electrodes were infinitely fast. Initially, on application of a dc potential difference, ΔV, the potential drop across the electrolyte, $\Delta\phi$, is equal to ΔV and the initial current I_0 yields the bulk conductivity ($I_0 = -\sigma\Delta V$). Thus, using the Nernst–Einstein relationship,

$$I_0/\Delta V = -F^2(D_+ + D_-)c_0 / RT$$

where c_0 is the initial concentration of the electrolyte and D_+ and D_- are the ionic diffusion coefficients.

After a time t, cations are consumed at the cathode and an equivalent number are produced at the anode; so a concentration gradient forms at each electrode. The changes in concentration at the electrode surfaces cause each electrode to develop a potential difference with respect to the initial electrode potential, one increasing, the other decreasing (Figure 9.10). At any time t, the cathode and anode salt concentrations are c_c and c_a, the sum of the potentials developed at the electrode is

$$\Delta E = RT/F \ln(c_a/c_c)$$

and the potential difference across the electrolyte is

$$\Delta\phi = \Delta V - \Delta E$$

As ΔE increases, therefore, the ionic migration currents decrease but ionic diffusion occurs because of the concentration gradient.

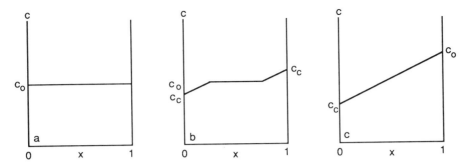

Figure 9.10. Schematic diagram of concentration profiles across the cell (**a**) at $t = 0$, (**b**) at $t > 0$ but before steady state is reached, and (**c**) at steady state.

At steady state, the cationic current I_+ can be given by

$$I_+ = -FD_+ \frac{dc}{dx} - \left(\frac{F^2D_+}{RT}\right) c \frac{d\phi}{dx}$$

        cation   cation

      diffusion   migration

where c is the electrolyte concentration at any distance x. Also,

$$I_+ = -2FD_+ \frac{dc}{dx}$$

As I_+ and D_+ are constant, the concentration gradient across the electrolyte is linear and so

$$I_+ = -2FD_+(c_a - c_c)$$

Providing the parameter $(c_a - c_c)/c_c$ is small,

$$(c_a + c_c/2)\ln c_a/c_c \sim (c_a - c_c) \tag{9.3}$$

and as $c_a + c_c = 2c_0$,

$$I_+ = -2D_+ c_0 \ln(c_a/c_c)$$

$$= (-2F^2D_+ \, c_0/RT)\Delta\phi$$

which gives an experimentally obtainable current–voltage relationship. As $\Delta V = 2\Delta\phi$,

$$I_+/\Delta V = -F^2 D_+ c_0/RT = -\sigma$$

which is the conductivity of the cell in the steady state and we have

$$I_+/I_0 = D_+/(D_+ + D_-) = t_+$$

This relationship is valid only for small concentration gradients and hence *small values of the applied potential*. From Figure 9.11, it is seen that provided $\Delta\phi$ is restricted to values below 10 mV, the ratio $c_0\ln(c_a/c_c)/(c_a-c_c)$, Eq. (9.3), does not deviate from unity by more than approximately 1%. Above 10 mV, the error increases rapidly and the relationships are no longer valid. This arises because potential difference is linearly related to c_a/c_c and current is linearly related to c_a-c_c, and as $\Delta\phi$ increases $c_0\ln c_a/c_c$ increasingly exceeds c_a-c_c.

Shriver and co-workers[100,101] used this relationship (i.e., on the assumption that only M^+ and X^- were mobile) to determine transport numbers of salts in various polymers. A polarizing voltage of 10 mV was applied and values of t_+ were found to range from 0.03 to 0.95.

This simple relationship, however, does not hold in cells where electrode kinetics are finite or where passivating layers (see Chapter 10) are present at the electrode surfaces.[99] It is unclear whether allowances were made for electrode effects, which would significantly affect the potential distribution and the current flowing in these cells. Similarly, a study was made of $LiClO_4$–PEO-based network polymers[102] to determine transference numbers using a combination of ac and dc techniques. Again it would appear no corrections were made for electrode reactions.

Evans et al.[103] described a technique in which a combination of dc and ac electrical polarizations are used to enable transference numbers to be determined, even when the diffusion coefficients of the ionic species are low or where precise correction for passivating layers or slow electrode kinetics is required. It can be shown that[99,103] under the same conditions as described earlier,

$$t_+ = \frac{I_+(\Delta V-I_0R_0)}{I_0(\Delta V-I_+R_s)}$$

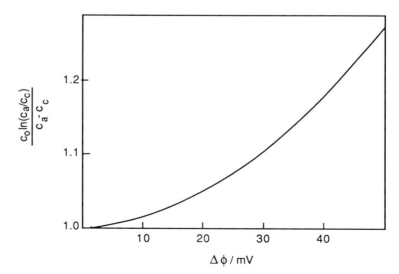

Figure 9.11. Ratio of $c_0\ln(c_a/c_c)$ to $c_a - c_c$ as a function of $\Delta\phi$. From Ref. 99.

where R_0 and R_s are the initial and steady-state currents. A value of $t_+ = 0.46 \pm 0.02$ was evaluated for $P(EO)_9LiCF_3SO_3$ using this method, whereas by equating T_+ with I_+/I_0, values between 0.48 and 0.61 were found for the same material.

Systems Containing Mobile Ion Pairs. Bruce and co-workers[59,60,104] developed the aforementioned theory to take account of ion pairing. It was assumed that ion pairs were in fast dynamic equilibrium with free ions and therefore a concentration gradient of MX would also be established. The steady-state current, I_+, is due to migration of M^+ cations in the electric field and diffusion of M^+ and MX down their concentration gradients. At steady state, for the anions,

$$\frac{F^2 D_-[X^-]}{RT}\frac{d\phi}{dx} = FD_-\frac{d[X^-]}{dx} + FD_0\frac{d[MX]}{dx}$$

anion	anion	ion pair
migration	diffusion	diffusion

where D_0 is the ion pair diffusion coefficient. For the cations,

$$I_+ = -FD_+\frac{d[M^+]}{dx} - FD_0\frac{d[MX]}{dx} - \frac{F^2 D_+[M^+]}{RT}\frac{d\phi}{dx}$$

$$= -FD_+([M^+]_a - [M^+]_c) - \frac{FKD_0(D_+ + D_-)}{D_-}([M^+]_a^2 - [M^+]_c^2)$$

where K is the equilibrium constant for ion association ($=[MX]/[M^+][X^-]$) and $[M^+]_a$ and $[M^+]_c$ represent the steady-state concentrations of M^+ adjacent to the anode and cathode, respectively. Also, as $[M^+] = [X^-]$ and $[MX] = K[M^+]^2$,

$$\Delta\phi = \frac{RT}{F}\left\{\ln\left(\frac{[M^+]_a}{[M^+]_c}\right) + \frac{2KD_0}{D_-}([M^+]_a - [M^+]_c)\right\}$$

and as before,

$$\Delta E = \frac{RT}{F}\ln\frac{[M^+]_a}{[M^+]_c}$$

$$\Delta V = \Delta\phi + \Delta E$$

$$= \frac{2RT}{F}\left\{\ln\frac{[M^-]_a}{[M^+]_c} + \frac{KD_0[M^+]_a - [M^+]_c}{D_-}\right\}$$

When the difference between the cation concentrations at the electrodes is small,

$$\ln\frac{[M^+]_a}{[M^+]_c} \sim \frac{2([M^+]_a - [M^+]_c)}{([M^+]_a + [M^+]_c)}$$

and

$$\frac{I_+}{\Delta V} = \frac{F^2 D_+ [M^+]_0}{RT} \left\{ 1 + \frac{KD_0 [M^+]_0}{D_+ (1 + KD_0 [M^+]_0 / D_-)} \right\} \tag{9.4}$$

where $[M^+]_0$ is the initial concentration of free ions through the electrolyte and is the average ionic concentration at the electrodes, $([M^+]_a + [M^+]_c)/2$. Where the ratio $I_+/\Delta V$ is independent of the applied voltage, it may be regarded as the effective conductivity at steady state, σ_{eff}. In all cases, a limiting value (for $\Delta V \rightarrow 0$) may be found. At low polarization values and for the mobility or concentration of ion pairs to be small in comparison with that of the anion, $KD_0 [M^+]_0 / D_- \ll 1$ and

$$\sigma_{eff} = F^2 D_+ [M^+]_0 / RT$$

which is the relationship derived previously. If ion pairs dominate, $KD_0 [M^+]_0 / D_- \gg 1$ and

$$\sigma_{eff} = F^2 [M^+]_0 (D_+ + D_-)/RT$$

which is the conductivity before polarization and implies that the flux of ion pairs down the concentration gradient is sufficient to balance the anion flux. Thus, if the total flux of M from anode to cathode is high, it is irrelevant which species carries the current. In the circumstances where M^+ is totally immobile, ion pairs would be the only means of transporting M. This has been considered by Ingram and co-workers.[105,106]

Unlike the situations for either completely dissociated electrolytes or immobile ion pairs, σ_{eff} is independent of the polarization voltage. The limits of validity of Eq. 9.4 and the conditions under which it is independent of ΔV have been determined by numerical simulations.[104] Steady-state current–potential linearity is expected over a wide potential range with the possibility of the linear range extending to as much as 10 V. Thus in electrolytes where ion pairing is significant, and the species are mobile, the relationship between applied voltage and steady-state current would be expected to remain linear for large values of applied voltage, whereas in systems whose mobile species are predominately free ions, significant deviations from linearity will be observed at applied voltages above 10 mV. Figure 9.12[60] shows a plot of reciprocal conductivity σ_{eff} against applied potential for the cell Li|PEO–LiClO$_4$|Li at steady state. At low concentrations of salt, deviations in the plot from linearity (i.e., variation in the value of σ_{eff} with applied voltage) occur at approximately 10 mV, and the linear range extends as the salt content increases. At steady state, increases in the bulk resistance from its nonpolarized value are found at low salt concentrations ($\sim$100:1).[107] This has been attributed to concentration differences between the electrodes. This effect will also give rise to deviations from linearity of the current–voltage plot and therefore both effects may limit the potential over which significant data can be obtained.

For the situation where neutral species are mobile, the ratio I_+/I_0 (σ_{eff}/σ) does not represent the transference number of species M because, by definition, it is a measure of the contribution of *charged* species to transport in an electric field.

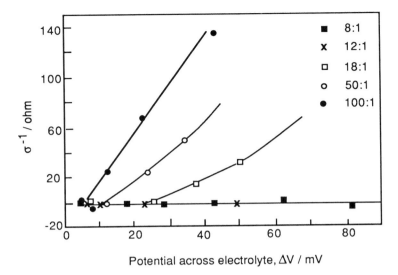

Figure 9.12. Variations in σ_{eff} with applied potential for various concentrations of $LiClO_4$ in PEO. The data are represented in such a way as to highlight the potentials at which deviations from linearity occur.

Neither the concentration nor the diffusion coefficient of the mobile species appears in the equation and so no information on transport parameters can be obtained. A better term for the ratio I_+/I_0 would be the "limiting current fraction"; "limiting" because it refers to the range of potential values over which I_+/I_0 is invariant, and "current fraction," as this indicates the largest proportion of the initial current that may pass through the electrolyte at steady state. Despite these factors, the fraction I_+/I_0 does have practical significance. In a battery, it is not important what species carries the current and it is useful, therefore, to have a comparative factor that relates to the transport of the electroactive constituent, M. Figure 9.13 shows the variation of this with concentration in the range 100:1 to 8:1 for PEO–$LiClO_4$ and PEO–$LiCF_3SO_3$ electrolytes. The ratio increases with decrease in salt content, with the value for the trifluoromethanesulfonate system approximately one-half that of perchlorate-based electrolytes at all concentrations.

Systems Containing Triple Ions. The additional effects of including higher associated species have also been considered.[107] The introduction of each new species, however, involves extra equilibrium constants and diffusion coefficients. By considering triple ions MX_2^- and M_2X^+, in addition to free ions and ion pairs, the equations become more complex. Taking probable combinations of these species simplifies the mathematics but at least 18 possible equations can be derived. It was concluded that, depending on the combination of species under consideration, the equations would simplify to either the fully dissociated situation or to the one for mobile ion pairs.

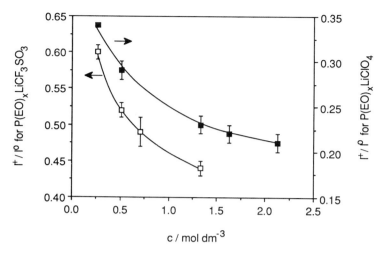

Figure 9.13. "Limiting current fraction" for PEO–LiCF$_3$SO$_3$ and PEO–LiClO$_4$ as a function of salt concentration. From Ref. 107.

To summarize, it would appear that some evidence is emerging to support the existence of mobile neutral ion pairs in particular concentration ranges in solid polymer electrolytes. If this is the case, then measurements of true transference numbers must be limited to methods such as the Hittorf method, which involve measuring only the transport of charged species in order to have meaningful, comparable data. For a battery, however, the total M transport is important, not the nature of the species carrying the current, provided the total flux of M from anode to cathode is high. In exceptional circumstances, ion pairs may provide the only means of transporting M through the cell. The preceding derivations have dealt with the simplest situations: free ions or ion pairs and free ions in an ideal electrolyte. The effects of nonideality and higher associated charged clusters have yet to be rigorously analyzed and the extent to which the preceding equations will deviate from the real situation remains unanswered.

References

1. H. Cheradame and J. F. Le Nest, in *Polymer Electrolyte Reviews—1* (J. R. MacCallum and C. A. Vincent, Eds.) Elsevier Applied Science, London (1987), p. 103.

2. J. F. Le Nest, H. Cheradame, and A. Gandini, *Solid State Ionics* **28–30** (1988), 1032.

3. J. F. Le Nest, F. Defendini, A. Gandini, H. Cheradame, and J. P. Cohen-Addad, *J. Power Sources* **20** (1987), 339.

4. M. Watanabe and N. Ogata, in *Polymer Electrolyte Reviews—1* (J. R. MacCallum and C. A. Vincent, Eds.), Elsevier Applied Science, London (1987), p. 39.

5. M. Watanabe, M. Itoh, K. Sanui, and N. Ogata, *Macromolecules* **20** (1987), 569.

6. M. Watanabe and N. Ogata, *Br. Polym. J.* **20** (1988), 181.

7. J. M. G. Cowie, A. J. S. Martin, and A. M. Firth, *Br. Polym. J.* **20** (1988), 247.

8. S. Bhattacharja, Diss. Abstr. Int B. **48** (1988), p. 3076.

9. R. M. Fuoss, *J. Am. Chem. Soc.* **57** (1935), 2604.

10. L. D. Pettit and S. Bruckenstein, *J. Am. Chem. Soc.* **88** (1966), 4783.

11. C. H. Porter and R. H. Boyd, *Macromolecules* **4** (1971), 589.

12. F. M. Gray, C. A. Vincent, and M. Kent, *J. Polym. Sci. Polym. Phys. Ed.* **27** (1989), 2011.

13. J. R. MacCallum, A. S. Tomlin, and C. A. Vincent, *Eur. Polym. J.* **22** (1986), 787.

14. L. M. Torell and S. Schantz, in *Polymer Electrolyte Reviews—2* (J. R. MacCallum and C. A. Vincent, Eds.), Elsevier Applied Science, London (1989), p. 1.

15. M. A. Ratner and A. Nitzan, *Faraday Discuss. Chem. Soc.* **88** (1989), 19.

16. M. C. Wintersgill, J. J. Fontanella, S. G. Greenbaum, and K. J. Adamic, *Br. Polym. J.* **20** (1988), 195.

17. H. Cheradame and P. Niddham-Mercier, *Faraday Discuss. Chem. Soc.* **88** (1989), 77.

18. J. F. Le Nest, H. Cheradame, and A. Gandini, *Solid State Ionics* **28–30** (1988), 1032.

19. G. G. Cameron, M. D. Ingram, and G. A. Sorrie, *J. Electroanal. Chem.* **198** (1986), 205.

20. P. G. Hall, G. R. Davies, I. M. Ward, and J. E. McIntyre, *Polym. Commun.* **27** (1986), 100.

21. F. M. Gray, *Solid State Ionics* **40/41** (1990), 637.

22. F. M. Gray, J. Shi, C. A. Vincent, and P. G. Bruce, *Phil. Mag.* (1991), in press.

23. G. G. Cameron, J. L. Harvie, M. D. Ingram, and G. A. Sorrie, *Br. Polym. J.* **20** (1988), 199.

24. G. G. Cameron, J. L. Harvie, M. D. Ingram, and G. A. Sorrie, *J. Chem. Soc. Faraday Trans. 1* **83** (1987), 3347.

25. B. L. Papke, M. A. Ratner, and D. F. Shriver, *J. Phys. Chem. Solids* **42** (1981), 493.

26. H. Sato and Y. Kusumoto, *Chem. Lett.* (1978), 635.

27. J. Eschmann, J. Strasser, M. Xu, Y. Okamoto, M. Eyring, and S. Petrucci, *J. Phys. Chem.* **94** (1990), 3908.

28. B. L. Papke, M. A. Ratner, and D. F. Shriver, *J. Electrochem. Soc.* **129** (1982), 1434.

29. R. Frech, J. Manning, D. Teeters, and B. E. Black, *Solid State Ionics* **28–30** (1988), 954.

30. B. L. Papke, R. Dupon, M. A. Ratner, and D. F. Shriver, *Solid State Ionics* **5** (1981), 685.

31. R. Dupon, B. L. Papke, M. A. Ratner, D. H. Whitmore, and D. F. Shriver, *J. Am. Chem. Soc.* **104** (1982), 6247.

32. M. Chabanel and Z. Wang, *J. Phys. Chem.* **88** (1984), 1441.

33. K. J. Maynard, D. E. Irish, E. Eyring, and M. Petrucci, *J. Phys. Chem.* **88** (1984), 729.

34. S. Schantz, L. Sandahl, L. Börjesson, L. M. Torell, and J. R. Stevens, *Solid State Ionics* **28–30** (1988), 1047.

35. M. Kakihana, S. Schantz, L. M. Torell, and L. Börjesson, in *Solid State Ionics,* Vol. **135** (G. Nazri, R. A. Huggins, and D. F. Shriver, Eds.), Materials Research Soc., Pittsburgh (1989), p. 351.

36. M. Kakihana, J. Sandahl, S. Schantz, and L. M. Torell, in *Second International Symposium on Polymer Electrolytes* (B. Scrosati, Ed.), Elsevier, London (1990), p. 1.

37. S. Schantz, L. M. Torell, and J. R. Stevens, *J. Appl. Phys.* **64** (1988), 2038.

38. M. Kakihana, S. Schantz, J. Stevens, and L. M. Torell, *Solid State Ionics* **40/41** (1990), 641.

39. R. Frech, J. Manning, and E. Hwang, *Second International Symposium on Polymer Electrolytes (ISPE-2), Siena, June 14–16* (1989), Extended Abstracts, p. 26.

40. F. M. Gray, *J. Poly. Sci. Polym. Lett. Ed.* (1991), in press.

41. D. E. Irish, in *Ionic Interactions: From Dilute Solutions to Fused Salts, Vol. 2* (S. Petrucci, Ed.), Academic Press, New York (1971), p. 187.

42. S. Clancy, D. F. Shriver, and L. A. Ochrymowycz, *Macromolecules* **19** (1986), 606.

43. M. Kakihana and S. Schantz, in *Second International Symposium on Polymer Electrolytes* (B. Scrosati, Ed.), Elsevier, London (1990), p. 23.

44. R. E. Hester and R. A. Plane, *Inorg. Chem.* **3** (1964), 769.

45. S. D. Ross, *Spectrochim. Acta* **18** (1962), 225.

46. M. G. Miles, G. Doyle, R. P. Cooney, and R. S. Tobias, *Spectrochim. Acta* **25A** (1969), 1515.

47. S. Schantz, M. Sandberg, and M. Kakihana, *Solid State Ionics* **40/41** (1990), 645.

48. S. G. Greenbaum, Y. S. Pak, M. C. Wintersgill, and J. J. Fontanella, *Solid State Ionics* **31** (1988), 241.

49. M. Spiro, in *Techniques of Chemistry*, Vol. 1, Pt. IIA (A. Weissberger and B. W. Rossiter, Eds.), Wiley, New York (1970).

50. M. Leveque, J. F. Le Nest, H. Cheradame, and A. Gandini, *Makromol. Chem. Rapid Commun.* **4** (1983), 497.

51. M. Leveque, J. F. Le Nest, A. Gandini, and H. Cheradame, *J. Power Sources* **14** (1985), 27.

52. P. Ferloni, G. Chiodelli, A. Magistris, and M. Sanesi, *Solid State Ionics* **18/19** (1986), 265.

53. J. E. Weston and B. C. H. Steele, *Solid State Ionics* **7** (1982), 81.

54. W. Gorecki, R. Andreani, C. Berthier, M. B. Armand, M. Mali, J. Roos, and D. Brinkmann, *Solid State Ionics* **18/19** (1986), 295.

• 55. A. Bouridah, F. Dalard, D. Deroo, and M. B. Armand, *Solid State Ionics* **18/19** (1986), 287.

56. P. M. Blonsky, D. F. Shriver, P. Austin, and H. R. Allcock, *Solid State Ionics* **18/19** (1986), 258.

57. J. Evans, C. A. Vincent, and P. G. Bruce, *Solid State Ionics* **28–30** (1988), 918.

58. D. F. Shriver, S. Clancy, P. M. Blonsky, and L. C. Hardy, in *Proceedings of the 6th Risø International Symposium on Metallurgy and Materials Science* (F. W. Poulsen, N. Hessel Andersen, K. Clausen, S. Skaarup, and O. T. Sørensen, Eds.), Risø National Laboratory, Roskilde (1985).

59. P. G. Bruce and C. A. Vincent, *Faraday Discuss. Chem. Soc.* **88** (1989), 43.

60. P. G. Bruce and C. A. Vincent, *Solid State Ionics* **40/41** (1990), 607.

61. A. V. Chadwick and M. R. Worboys, in *Polymer Electrolyte Reviews—1* (J. R. MacCallum and C. A. Vincent, Eds.), Elsevier Applied Science, London (1987), p. 275.

62. A. V. Chadwick, J. H. Strange, and M. R. Worboys, *Solid State Ionics* **9/10** (1983), 1155.

63. C. Bridges, A. V. Chadwick, and M. R. Worboys, *Br. Polym. J.* **20** (1988), 207.

64. A. A. Al-Mudaris and A. V. Chadwick, *Br. Polym. J.* **20** (1988), 213.

65. A. V. Chadwick, A. A. Al-Mudaris, and C. Bridges, *Polym. Prepr.* **30** No. 1 (1989), 418.

66. M. Minier, C. Berthier, and W. Gorecki, *J. Phys. (Paris)* **45** (1984), 739.

67. S. Bhattacharja, S. W. Smoot, and D. H. Whitmore, *Solid State Ionics* **18/19** (1986) 306.

68. S. E. Lindsey, D. H. Whitmore, W. P. Halperin, and J. M. Torkelson, *Polym. Prepr.* **30** No. 1 (1989), 442.

69. W. Gorecki, R. Andreani, C. Berthier, M. Armand, M. Mali, J. Roos, and D. Brinkman, *Solid State Ionics* **18/19** (1986), 295.

70. M. Mali, J. Roos, and D. Brinkmann, in *Proceedings, XXII Congress Ampère* (K. A. Muller, R. Kind, and J. Roos, Eds.), Zurich (1984).

71. W. Gorecki, P. Donoso, C. Berthier, M. Mali, J. Roos, D. Brinkmann, and M. B. Armand, *Solid State Ionics* **28–30** (1988), 1018.

72. M. Armand, *Faraday Discuss. Chem. Soc.* **88** (1989), 65.

73. M. R. Worboys, Ph.D. thesis, University of Kent at Canterbury (1985).

74. F. L. Tanzella, W. Bailey, D. Frydrych, G. C. Farrington, and H. Story, *Solid State Ionics* **5** (1981), 681.

75. A. Killis, J. F. Le Nest, A. Gandini, H. Cheradame, and J-P. Cohen-Addad, *Polym. Bull.* **6** (1982), 351.

76. M. C. Wintersgill, J. J. Fontanella, J. P. Calame, S. G. Greenbaum, and C. G. Andeen, *J. Electrochem. Soc.* **131** (1984), 2208.

77. M. C. Wintersgill, J. J. Fontanella, J. P. Calame, M. K. Smith, T. B. Jones, S. G. Greenbaum, K. J. Adamic, A. N. Shetty, and C. G. Andeen, *Solid State Ionics* **18/19** (1986), 326.

78. M. McLin and C. A. Angell, *Polym. Prepr.* **30** No. 1 (1989), 439.

79. P. R. Sørensen and T. Jacobsen, *Solid State Ionics* **9/10** (1983), 1147.

80. M. Watanabe, K. Sanui, N. Ogata, T. Kobayashi, and Z. Ohtaki, *J. Appl. Phys.* **57** (1985), 123.

81. M. Watanabe, S. Nagano, K. Sanui, and N. Ogata, *Solid State Ionics* **18/19** (1986), 338.

82. M. Watanabe, M. Rikukawa, K. Sanui, and N. Ogata, *J. Appl. Phys.* **58** (1985), 736.

83. M. Watanabe, M. C. Longmire, and R. W. Murray, *J. Phys. Chem.* **94** (1990), 2614.

84. V. Cammarata, D. R. Talham, R. M. Crooks, and M. S. Wrighton, *J. Phys. Chem.* **94** (1990), 2680.

85. H. Cheradame, A. Killis, L. Lestel, and S. Boileau, *Polym. Prepr.* **30** No. 1 (1989), 420.

86. A. Bouridah, F. Dalard, and M. B. Armand, *Solid State Ionics* **28–30** (1988), 950.

87. D. A. MacInnes and B. R. Ray, *J. Am. Chem. Soc.* **71** (1949), 2987.

88. D. A. MacInnes and B. R. Ray, *J. Phys. Chem.* **61** (1957), 657.

89. P. R. Sørensen and T. Jacobsen, *Electrochim. Acta* **27** (1982), 1671.

90. J. R. MacDonald, *J. Chem. Phys.* **58** (1973), 4982.

91. J. R. MacDonald, *J. Chem. Phys.* **61** (1974), 3977.

92. L. L. Yang, R. Huq, and G. C. Farrington, *Solid State Ionics* **18/19** (1986), 291.

93. L. L. Yang, A. R. McGhie, and G. C. Farrington, *J. Electrochem. Soc.* **133** (1986), 1380.

94. J. E. Weston and B. C. H. Steele, *Solid State Ionics* **7** (1982), 81.

95. P. Ferloni, G. Chiodelli, A. Magistris, and M. Sanesi, *Solid State Ionics* **18/19** (1986), 265.

96. D. Fauteux, *Solid State Ionics* **17** (1985), 133.

97. J. W. Lorimer, *J. Power Sources* **26** (1989), 491.

98. R. Pollard and T. Compte, *J. Electrochem. Soc.* **136** (1989), 3734.

99. P. G. Bruce and C. A. Vincent, *J. Electroanal. Chem.* **225** (1987), 1.

100. P. M. Blonsky, D. F. Shriver, P. Austin, and H. R. Allcock, *Solid State Ionics* **18/19** (1986), 258.

101. D. F. Shriver, S. Clancy, P. M. Blonsky, and L. C. Hardy, in *Proceedings, 6th Risø International Symposium on Metallurgy and Materials Science* (F. W. Poulsen, N. Hassel Andersen, K. Clausen, S. Skaarup, and O. T. Sørensen, Eds.), Risø National Lab., Roskilde (1985).

102. M. Watanabe, S. Nagano, K. Sanui, and N. Ogata, *Solid State Ionics* **28–30** (1988), 911.

103. J. Evans, C. A. Vincent, and P. G. Bruce, *Polymer* **28** (1987), 2324.

104. P. G. Bruce, M. D. Hardgrave, and C. A. Vincent, *J. Electroanal. Chem.* **271** (1989), 27.

105. G. G. Cameron, J. L. Harvie, and M. D. Ingram, *Solid State Ionics* **34** (1989), 65.

106. G. G. Cameron, M. D. Ingram, and J. L. Harvie, *Faraday Discuss. Chem. Soc.* **88** (1989), 55.

107. M. D. Hardgrave, Ph.D. thesis, St. Andrews (1990).

CHAPTER 10

The Electrode–Electrolyte Interface

The importance of understanding the electrode–electrolyte interface when evaluating an electrochemical system is obvious. At the interface, charge distribution along with physical properties—crystallographic, mechanical, compositional, and particularly electrical—can reduce the overall electrical conductivity of a system. Despite this, relatively few studies have been carried out, confined in the main to the lithium–polymer electrolyte interface with a little information on the positive electrode–polymer interface.

10.1. The Lithium–Polymer Electrolyte Interface

The lithium–polymer electrolyte interface has been studied using ac impedance spectroscopy,[1–6] which clearly shows the presence of a resistive layer. The impedance is time dependent but usually reaches a constant value after a number of hours of contact between lithium and the electrolyte[1,3] (Figure 10.1). It was initially thought that this layer was due to trace quantities of water reacting with the metal and forming an oxide layer, but in a study by Fauteux[1] on scrupulously dry films it was observed that the formation of a passivating film on the lithium electrode did not appear to be related simply to the presence of residual water. Rather, it was postulated that two separate film formation reactions occurred simultaneously and were competitive. One involved reaction of lithium with catalytic residues in the polymer (Ca and Si contents 6000 and 8000 ppm, respectively) and dominated at low salt content. The other was attributed to reaction of Li with $LiCF_3SO_3$ with the formation of LiF. Because of the strength and stability of the C—F bond in comparison with C—S an alternative process has been proposed[7]:

$$LiCF_3SO_3 + Li(s) \rightarrow 2Li^+ + SO_3^{2-} + CF_3^{\cdot}$$

The radicals would subsequently form CF_3H by hydrogen abstraction from the polymer backbone, giving lithium sulfite as the passivating film. The electrolyte salt and its purity have a determining role in the kinetics of the passivation process and the nature of the layer formed, but the thickness (i.e., resistance) of the layer is found to be salt concentration independent.[8] Hiratani[6] interpreted this low-frequen-

215

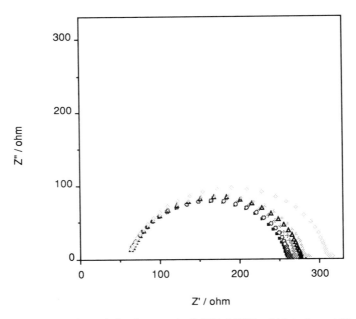

Figure 10.1. Growth of a resistive layer at the $P(EO)_8LiClO_4$–Li interface at 95.5°C: (■) t = 0, (□) t = 0.5 h, (+) t = 2.5 h, (△) t = 4 h, (▲) t = 5 h, (◇) t = 22 h.

cy semicircle as a combination of interfacial resistance and charge transfer resistance. Variations in geometric capacitance with temperature cycling were also reported and were interpreted in terms of a reduction in contact area brought about by disconnections occurring during cooling of the sample. The reactivity of the Na:PEO–NaClO$_4$ interface [PEO = poly(ethylene oxide)] has been studied by West et al.[9] using ac spectroscopy. Initially, the cell impedance is dominated by a large interfacial impedance and two relaxation arcs are identified (Figure 10.2). The impedance is time dependent as can be seen from the low-frequency impedance trend. After 24 h the interfacial impedance drops considerably and three distinct arcs are observed on the plot. These have been ascribed to (I) resistance of an ionically conducting film on the sodium surface, (II) transport in an inhomogeneous porous layer of corrosion products dispersed in the electrolyte, and (III) a diffusion arc caused by simultaneous transport of anions and cations in the bulk. Whatever the nature of films formed at the metal–polymer electrolyte interface, it has a profound effect on the behavior of electrochemical cells and should always be taken into consideration when interpreting, for example, dc polarization data for transference number measurements as described in Chapter 9.

Sequeira and Hooper[10] and Bonino et al.[11,12] used dc micropolarization methods to determine whether the electrode process at a PEO–LiCF$_3$SO$_3$ interface could be modeled by a simple charge-transfer reaction. Variation in the exchange current density, i_0, was found in the former study, the value increasing with applied pres-

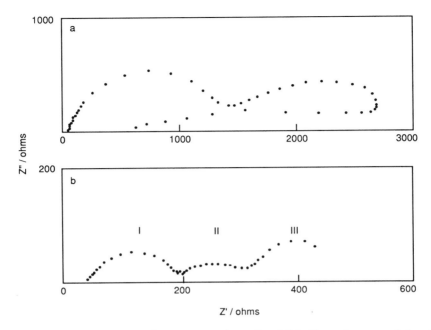

Figure 10.2. Impedance of a Na|PEO$_{12}$NaClO$_4$|Na cell at 80°C. Electrode area = 0.7 cm^2, electrolyte thickness = 100 μm. (**a**) Immediately after cell assembly, (**b**) after equilibration. From Ref. 9. Reproduced by permission of the SCI.

sure, indicating a variable contact area. The low-pressure values were close to those reported by Bonino et al.[11] These values did not, however, agree with the data of Fauteux et al.[8] for a high-purity PEO electrolyte, PEO–LiCF$_3$SO$_3$.[8] The enthalpy of activation at zero polarization was calculated by Sequeira and Hooper to be approximately 75 to 100 kJ mol^{-1}, independent of concentration, although high-pressure, high-temperature values of approximately 29 kJ mol^{-1} were found. These values were higher than expected for metal dissolution and deposition (~42 kJ mol^{-1}), suggesting some other interfacial process. Fauteux[1] examined the experimental data of Sequeira and Hooper and concluded that the low-field simplification to the Butler–Volmer equation used,

$$i = i_0 F\eta/RT$$

where η is the overpotential and other symbols take their usual meaning, could not be used because the low-field condition, $F\eta/RT \ll 1$, was not fulfilled. Despite this, linear dependence of the current and overvoltage was observed and was associated with ohmic resistance behavior of the passivating layer. The change in activation energy at high temperature was explained by Fauteux as resulting from calculations on a still-growing passivating layer.

Fauteux et al.[8] looked at the interfacial layer behavior between the polymer electrolyte and Li–Al and Li–Sb electrodes. The alloys showed behavior similar to that of Li. Lithium–aluminum electrodes in polymer electrolyte-based cells have also been studied by Owen et al.[13] The kinetics of charging the Al foil to form a 20-μm layer of Li–Al were found to be limited predominantly by the overpotential for growth of the βLi–Al phase. The diffusion coefficient of lithium in this phase is high enough not to limit the charging rate. The electroplating of lithium onto a number of other electronically conducting substrates and its subsequent stripping have also been studied. Platinum, nickel, stainless steel, and vitreous carbon have been investigated.[11,12,14] It was reported that virtually every metal forms at least dilute alloy phases with lithium, making the reversibility of initial lithium plating poor. For example, Figure 10.3 shows the cyclic voltammogram for lithium plating and stripping on nickel at 80°C. On the cathodic scan, the voltage required to deposit the metal on the nickel surface is some 250 mV more negative than that on lithium itself. Once the monolayer is formed, however, further plating can occur at any potential below zero volts so that a "reverse peak" is formed on commencing the anodic scan. At positive potentials, lithium is reoxidized. At high scan rates, good cyclability was found, but at lower ratios recovery of lithium decreased considerably. These results suggest again that freshly deposited lithium undergoes some passivation reaction. Reversibility of the plating process was highly dependent on the nature of the substrate; for example, it was very poor for vitreous carbon possibly because of slow diffusion of lithium into the bulk of the carbon. Investigations of the time dependence of the dc potential of lithium showed there to be an

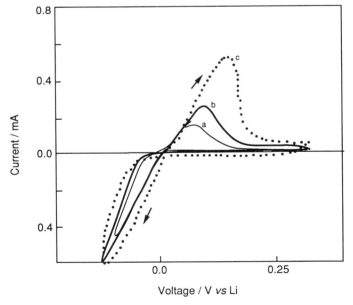

Figure 10.3. Cyclic voltammogram of the plating–stripping process. Nickel substrate, $P(EO)_9LiCF_3SO_3$, at 80°C for (**a**) 1 mV s^{-1}, (**b**) 5 mV s^{-1}, and (**c**) 20 mV s^{-1}. From Ref. 12.

interplay of a number of processes in which the lithium interacts with both electrode material and components of the electrolyte. The length of time the cell voltage remained at zero following a plating pulse was found to be dependent on the amount of lithium plated, on the nature of the substrate, and on the previous plating history of the cell, and it was proposed that alloy formation was principally responsible for this; however, modifications of the lithium surface by reaction with the electrolyte should also be considered as a determining factor.[15]

10.2. Electrochemical Stability

A useful polymer electrolyte, particularly for electrochemical power sources, must show stability to both oxidation and reduction. Generally what is important is the practical kinetic-limited stability rather than the thermodynamically based absolute stability. The formation of passivating layers on the surface of lithium electrodes modifies the electrochemical behavior of the electrode, for example, by introducing a "voltage delay"; these layers may also be beneficial in helping to extend the shelf life of the cell. The ether linkage in polyethers such as PEO is particularly favorable with respect to stability as the C—O bond is one of the least reactive (360 kJ mol^{-1}) in organic chemistry. A wide range of polymer electrolytes other than those based on linear PEO have now been synthesized (Chapter 6) with a view toward producing higher-conducting and/or more mechanically stable systems. This usually results in the inclusion of further organic functional groups which could seriously affect electrochemical stability and render the materials useless for practical devices. Of all the new materials currently under study, the electrochemical stability of only one system,[16] a poly(dimethyl siloxane-grafted ethylene oxide) copolymer crosslinked by an aliphatic urethane linkage, has been considered. It was reported to have a stability range of more than 3 V even though unreacted isocyanate groups were shown to be very unstable.

Armand and co-workers examined the redox stability domain of a number of electrolytes including $P(EO)_{4.5}LiI$, $P(EO)_{4.5}NaCF_3SO_3$, $P(EO)_{4.5}LiClO_4$ $P(EO)_8LiClO_4$, and $P(EO)_{4.5}LiCF_3SO_3$ by cyclic voltammetry using a three-electrode cell with a Ag/Ag_3SI reference electrode and platinum working electrode, examples of which are given in Figure 10.4.[17-19] The domain is limited on the cathodic side by the lithium deposition–dissolution process, a process that is rather more complex than $Li^+ + e^- \rightarrow Li(s)$ because of the formation of passivating layers and low-concentration alloys which have been described earlier. On the anodic side, the electrochemical window is limited by the irreversible oxidation of the anion. For PEO–LiI systems, a window of 2.8 V was reported, and for PEO–$NaCF_3SO_3$ a value just under 4.0 V was found. In most instances the stability window appears to be broad enough to permit the use of most common electrode couples for lithium batteries; however, certain anions, typically SCN^-, NO_3^-, and $B(C_6H_5)_4^-$, were reported to undergo reductive cleavage prior to metal deposition, for example,

$$SCN^- + 2e^- \rightarrow S^{2-} + CN^-$$

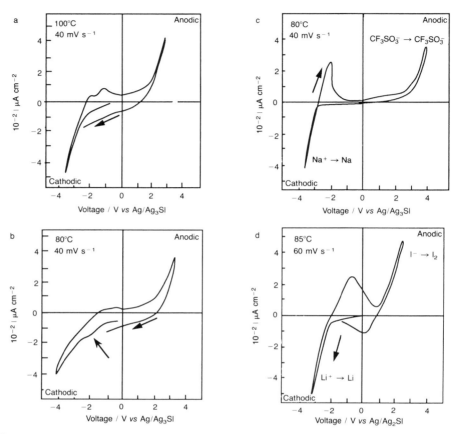

Figure 10.4. Cyclic voltammogram for (**a**) P(EO)$_8$LiClO$_4$,[18] (**b**) P(EO)$_{4.5}$LiCF$_3$SO$_3$[18] (arrow indicates possible anion reduction), (**c**) P(EO)$_{4.5}$NaCF$_3$SO$_3$,[22] and (**d**) P(EO)$_{4.5}$LiI[22] at 80 to 100°C. Scan rate = 40 to 60 mV s^{-1}, reference = Ag/Ag$_3$SI.

The trifluoromethane sulfonate ion appears to be more susceptible to reductive cleavage than perchlorate as suggested by the shoulder on the cyclic voltammogram at approximately 1.7 V. This has serious implications for the use of some salts in rechargeable lithium batteries as decomposition may occur on charging. Hooper and co-workers[20] also examined the electrochemical stability of P(EO)$_9$LiCF$_3$SO$_3$, but rather than assuming reduction of the anion in the cathodic region, they have interpreted it as a reduction of residual water within the electrolyte.

Most PEO electrolytes based on oxyanions show an oxidation peak at 4.0 to 4.5 V versus Li$^+$/Li. It has been proposed that the perchlorate anion is initially oxidized to form a radical

$$ClO_4^- - e^- \rightarrow ClO_4 \cdot$$

but it is unclear whether disproportionation

$$ClO_4\cdot \rightarrow ClO_2 + O_2$$

and/or hydrogen abstraction from the polymer backbone

$$ClO_4^- + H-\overset{|}{\underset{|}{C}}-H \longrightarrow HClO_4 + \cdot\, \overset{|}{\underset{|}{C}}-H$$

leads to the subsequent degradation products.[18] $CF_3SO_3^-$ oxidizes initially to form the radical. Because trifluoromethanesulfonic acid is a very strong acid, hydrogen abstraction is unlikely and therefore the relatively weak C—S bond is liable to attack:

$$CF_3SO_3\cdot \rightarrow CF_3\cdot + SO_3$$

At this point the $CF_3\cdot$ radical may attack the polymer to give degradation products, limited possibly by the rate of $CF_3\cdot$ dimerization to give unreactive C_2F_6 gas.

The stability of the polymer electrolyte shows significant temperature dependence. The stability window of $P(EO)_9LiCF_3SO_3$ is reduced from 3.8 V at 100°C to 3.3 V at 140°C to 1.9 V at 170°C. It may be that this behavioral change is related to the melting of crystalline $P(EO)_{3.5}LiCF_3SO_3$ around 140°C. For some practical applications temperature control may be necessary to prevent overheating and subsequent irreversible degradation.

Nonalkali metal systems have also been studied. Abrantes et al.[21] reported on the stability domains of $P(EO)_8AgClO_4$, $P(EO)_{12}ZnCl_2$, $P(EO)_8(CuClO_4)_2$, and $P(EO)_8CuCl_2$. The stability domains were limited by discharge of cations on the cathodic side and by Cl^- or ClO_4^- oxidation on the anodic side. For PEO–$Cu(ClO_4)_2$, the decomposition voltage was shown to fall rapidly with temperature. No evidence for the rather complex electrode processes observed for PEO–$Cu(CF_3SO_3)_2$ systems discussed in Chapter 7 was apparent here, which may result from residual water or solvent in the latter electrolytes. Alkaline earth salt polymer electrolytes have been studied but their redox stability has received little attention. Armand[22,23] has reported that magnesium does not appear to be electroactive over the whole potential range because of strong polymer–cation interactions (see Chapter 7). By comparison, as demonstrated in Figure 10.5, Ba^{2+} shows a reduction peak in a mixed PEO–$LiClO_4$–$Ba(ClO_4)_2$ system, presumably because of its larger size and weaker solvation interaction.

10.3. Intercalation

An intercalation material may be defined as a solid that can interact with cations and electrons from external sources, generating a new solid in which structural elements of the initial solid are maintained. Such elements may be tri-, bi-, or unidimensional structures in which cations become intercalated, usually without their coordination

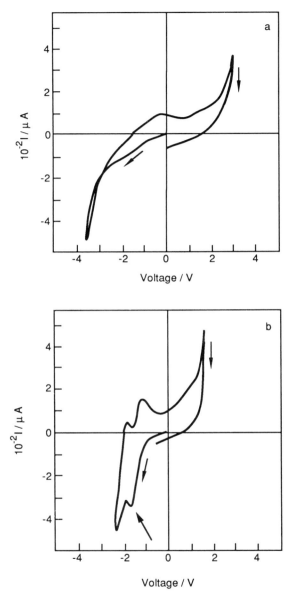

Figure 10.5. Cyclic voltammogram for (**a**) $P(EO)_8-(0.9)LiClO_4-(0.1)Mg(ClO_4)_2$, Pt electrode; (**b**) $P(EO)_8-(0.9)LiClO_4-(0.1)Ba(ClO_4)_2$ (arrow shows Ba^0 deposition peak), at 85°C. Scan rate = 40 mV s^{-1}, reference = Ag/Ag_3SI. From Ref. 17.

sphere of solvent. Compounds capable of intercalating lithium with only minor structural changes

$$MX_n + aLi^+ + ae^- \rightarrow Li_aMX_n$$

are of considerable interest as cathode materials in lithium batteries. Intercalation compounds are of particular commercial interest because they offer potentially high gravimetric and volumetric energy and power densities in lithium cells. They are also of great relevance in electrochromic devices. A wide variety of inorganic compounds with either two-dimensional van der Waals-bonded layer structures or three-dimensional framework tunnel structures undergo topotactic reactions with lithium.[24] A number of chalcogenides such as TiS_2 appear to be promising candidates. Murphy et al.[25] illustrated the importance of both electronic conductivity and unit cell volume to the topochemical incorporation of lithium by transition metal oxides with rutile-related structures. It was suggested that the unit cell volumes of the first-row transition metal rutiles are too small to incorporate lithium at significant rates. Improvement in lithium insertion by anatase TiO_2,[26] $\gamma-MnO_2$,[27] and several chromium oxides[28] compared with their rutile forms may be attributed to the lower density of the former.

The reversibility, stability, and absence of co-intercalation of the solvent (polymer) are among the potential advantages of using these materials in polymer electrolyte-based cells. The latter phenomenon is often seen for liquid electrolytes and two-dimensional intercalating structures that are not sterically selective and occurs when the cation entering the electrode material is too strongly solvated to lose its solvation sheath. For example, it has been observed for propylene carbonate-based electrolytes in contact with the chalcogenides TiS_2, ZrS_2, and TaS_2.[29] Chabre et al.[30] have used to advantage the absence of side reactions in polymer electrolyte cells to set up ultraslow scan voltammetry (10^{-5} V s^{-1}). This technique allows direct differentiation between the sites according to their energies in the intercalation compound. Figure 10.6 shows an ultraslow scan voltammetry trace for lithium insertion into $ZrSe_{1.915}$ using a $P(EO)_8LiClO_4$ electrolyte. The first oxidation peak at 2.25 V has been interpreted as the preferential intercalation of lithium on sites bounded to selenium vacancies which irreversibly modify the structure. On further intercalation, a semiconducting, reversible intercalating state is observed that becomes metallic at higher lithium content.

As far as polymer electrolytes are concerned, TiS_2 and V_6O_{13} have received the most attention as cathodic materials. $Li_{x+1}V_3O_8$ has also been studied. WO_3 in particular has been investigated as an electrode material for electrochromic display purposes. TiS_2 has a sandwich structure comprising TiS_6 octahedra sharing edges to give two-dimentsional S—Ti—S sheets. Between these sheets are vacant octahedra and tetrahedral sites into which small cations such as Li^+ can diffuse. For larger ions, the S—TI—S sheets shift relative to one another, giving trigonal prismatic sites between the sheets. The characteristics of TiS_2/polymer electrolyte interfaces and the kinetics of the electrochemical intercalation process have been examined by ac complex impedance spectra and cyclic voltammetry.[31] The impedance diagram is shown in Figure 10.7. The high-frequency semicircle may be associated with the

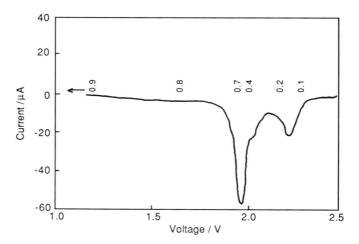

Figure 10.6. Ultraslow scan voltammetry trace for lithium insertion into $ZnSe_{1.915}$. Electrolyte = $P(EO)_8LiClO_4$, electrode capacity = 1.2 mAh, reference = Li^0, $T = 100°C$; scan rate = 10 mV h^{-1}. Arrow indicates increasing theoretical lithium content. From Ref. 30.

charge transfer process and the Warburg impedance to diffusion of Li into TiS_2. Good cyclability has been confirmed by cyclic voltammetry. Steele et al.[3] also investigated the cycling characteristics of Li|PEO–LiCF$_3$SO$_3$|Li cells. Capacities exhibited by the positive electrode at 110°C were 40 to 50% of theoretical values at C/10 rates and fell to 25% after 25 cycles. Although this is inferior to capacities of cells containing organic liquid electrolytes, the latter exhibited a serious loss of capacity as a result of problems associated with the reactivity at the lithium elec-

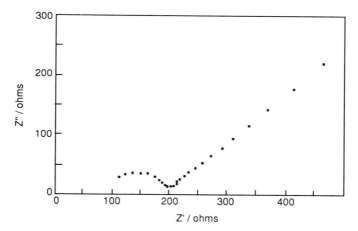

Figure 10.7. Impedance plot of the $TiS_2/P(EO)_9LiCF_3SO_3$ interface at 100°C. Lithium reference electrode. From Ref. 31. Reproduced by permission of the SCI.

trode. This loss was much less severe for polymeric systems. Johnson and Worrell[32] studied alkali metal interactions in TiS_2 in a polymer electrolyte cell. Open-circuit voltages were measured on a $Li|P(EO)_{4.5}LiC_2O_2F_3|Li_xTiS_2$ cell. Electromotive force results were comparable to those of a similar cell using dioxolane–$LiPF_6$ electrolyte. This varied from approximately 1.95 V for $x = 1$ to approximately 2.4 V for $x = 0.2$. Patrick et al.[33] recorded the open-circuit voltage for TiS_2 and various oxides using alkali earth metal anodes and alkali earth metal thiocyanates or perchlorates. These are listed in Table 10.1.

Scrosati[31] has described the properties of $Li_{1+x}V_3O_8$ as a lithium intercalation electrode in contact with a PEO–$LiCF_3SO_3$ electrolyte. The basic structural elements of this material are octahedra and trigonal bipyramids arranged to form puckered layers with Li^+ ions situated inbetween (Figure 10.8). The unit cell structure allows an excess of lithium to be accommodated with minimal structural distortion.[34] The maximum lithium uptake corresponds to $x \sim 3$, although above $x = 0.15$ a structural rearrangement occurs. X-ray analysis shows this to be highly reversible. The impedance diagram in Figure 10.9 is for a $Li_{1+x}V_3O_8|$PEO–$LiCF_3SO_3$ interface. The Warburg line again suggests diffusion-controlled kinetics. Fast lithium diffusion, at least at the initial stages of the electrochemical process, is expected for this system. Two semicircles are apparent in the impedance plots; the lower-frequency one is probably due to an interfacial layer, which is discussed later with reference to the V_6O_{13} interface. The reversibility of this electrode material in polymer electrolyte cells is demonstrated by cyclic voltammetry. Faster diffusion in PEO–$LiClO_4$ to PEO–$LiCF_3SO_3$ systems was implied by the resolved features in

Table 10.1. Open-Circuit Voltage for Cells with Various Anodes and Cathodes for Mg-Based Polymer Electrolytes

Cathode	Anode	Open-Circuit Voltage (V)
TiS_2	Mg	1.70
	Ca	2.28
	Zn	0.87
	Al	0.87
V_6O_{13}	Mg	2.00
	Ca	2.75
	Zn	1.35
	Al	1.25
MnO_2	Mg	2.00
	Ca	2.50
	Zn	1.23
	Al	1.25
NiO_2	Mg	1.50
CoO_2	Mg	1.65
MoO_2	Mg	1.75
MoO_3	Mg	1.75
V_2O_5	Mg	1.45
WO_3	Mg	1.80

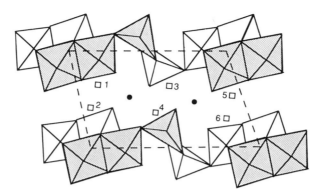

Figure 10.8. Structure of $Li_{1+x}V_3O_8$ projected onto (0,1,0) showing both octahedral and trigonal bipyramids. (●) Octahedral Li, (□) tetrahedral Li.

the voltammograms for the former, which arise from the inequivalence of the tetrahedral sites occupied by the Li^+ ions in the vanadium bronze structure.

Because of its high affinity for lithium insertion, V_6O_{13} has received widespread attention as a possible electrode material. In addition, the theoretical energy density of the Li/V_6O_{13} couple is 890 Wh kg^{-1} which is considerably higher when compared with other intercalation cathodes such as TiS_2. Despite this, it does not have the same rate capability or conductivity so, under load, its energy density is no higher. The monoclinic structure contains chains of tricapped perovskite cavities interconnected in the crystallographic (0,1,0) direction by square open faces. The chains are pairwise connected in the (0,0,1) direction by other open faces. Li^+ ions

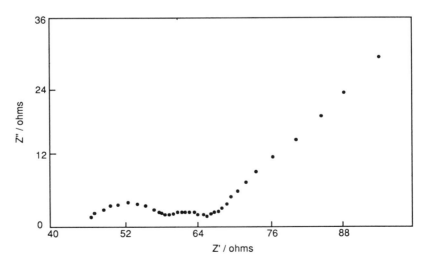

Figure 10.9. Impedance diagram of $Li_{1+x}V_3O_8/P(EO)_9LiCF_3SO_3$ interface at 100°C. Lithium reference electrode. From Ref. 31. Reproduced by permission of the SCI.

can be accommodated at oxygen-coordinated sites in the cavities. The maximum Li^+ uptake corresponds to a reduction of vanadium to the trivalent state, that is, to the composition $Li_8V_6O_{13}$.

A few studies of the V_6O_{13}/polymer electrolyte interface have been carried out using variable-frequency ac impedance methods.[3,35,36] Steele et al.[3] analyzed the ac response of V_6O_{13} electrodes at temperatures up to 146°C. They suggested that appreciable electrode resistance was likely to limit the use of this material as an electrode to temperatures above 140°C. At lower temperatures, large interfacial impedances were responsible for limiting the available capacity to 10 to 20% of the theoretical value. It was thought that this poor performance may be associated in part with the electrode material comprising pure V_6O_{13} powder rather than a composite which would incorporate a proportion of the electrolyte material. Bruce et al. have studied the V_6O_{13} interface with PEO–LiCF$_3$SO$_3$ and PEO–LiClO$_4$[35,36] using three-electrode cells. Electrodes of pure V_6O_{13} pressed powder, a composite containing V_6O_{13} (45%), polymer electrolyte (50%), and acetylene black (5%) by volume along with $Li_xV_6O_{13}$, where the lithium had been previously intercalated, were studied. The ac responses at both solid and liquid polymer electrolyte–cathode interfaces were evaluated. All samples exhibited the same trends in behavior. Figures 10.10 and 10.11 show the impedance behavior for polycrystalline and composite cathode material in contact with the solid polymer electrolyte. Initially, only one semicircle was evident but with time a second semicircle evolved, and both continued to grow, the latter at a slower rate. In the case of the composite cathode, the evolution of the second semicircle was distinctly slower. It was noted, however, that at high lithium content in $Li_xV_6O_{13}$ (above $x = 6$) the second semicircle grew more rapidly with time. Variations in the electrolyte or its composition had little effect on the impedance diagrams. The equivalent circuit for these ac responses is given in Figure 10.12 where Z_d represents the diffusional impedance that must exist at lowest frequencies. R_b is the bulk resistance associated with the intercept at highest frequencies on the Z' axis, R_{SL} and C_{SL} are associated with the high-frequency semicircle, and R_e and C_e are associated with the low-frequency semicircle. It was suggested that the low-frequency semicircle may be associated with processes occurring directly at the electrode surface, whereas the other was due to a layer adjacent to it. The complex impedance plane plots for P(EO)$_{20}$LiCF$_3$SO$_3$|TiS$_2$ and P(EO)$_{20}$LiCF$_3$SO$_3$|WO$_3$ obtained in this study again showed two semicircles that grew with time. This contrasts with Figure 10.7 which shows the complex plane plot for P(EO)$_9$LiCF$_3$SO$_3$|TiS$_2$ and reveals only one semicircle[31]; however, the cell history is not known in this instance. The V_6O_{13} electrodes in contact with liquid and solid electrolytes behave similarly except when the liquid polymer is subjected to a purification procedure. No interface impedance is seen in this instance.[37] It may be concluded that interfacial impedances arise from reactions with polymer impurities (different in solid and liquid systems) or with the lithium salt rather than the polymer itself. These impedances were found to grow until they were significantly larger than the bulk resistance, and therefore methods of minimizing this growth are advantageous to cell development.

Armand and co-workers[38] studied morphological changes of cathodic materials

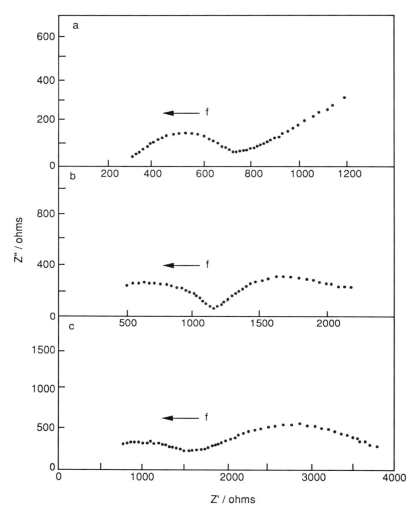

Figure 10.10. $P(EO)_{20}LiCF_3SO_3/V_6O_{13}$ polycrystalline cathode interface at 100°C at (**a**) 2 h, (**b**) 200 h, and (**c**) 320 h after cell assembly. From Ref. 35.

during discharge of polymer electrolyte-based lithium batteries by *in situ* scanning electron microscopy. TiS_2 and V_6O_{13} were investigated as insertion compounds and FeS as an example of a displacement material, that is, $2Li + FeS \rightarrow Li_2S + Fe$. It was found that TiS_2, which undergoes bidimensional intercalation and is a single-phase material, remains unchanged during discharge, although in overdischarge, that is, Ti reduced beyond the $3+$ valence state, the particles swell and crack. V_6O_{13}, which is a one-dimensional two-phase insertion compound, cracked during Li insertion, with the cracks widening on second discharge. Macroscopically the

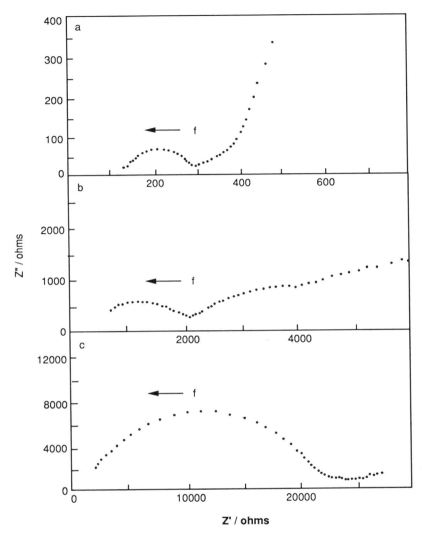

Figure 10.11. $P(EO)_{20}LiCF_3SO_3/V_6O_{13}$ composite cathode interface at 100°C at (**a**) 2 h, (**b**) 90 h, and (**c**) 220 h after cell assembly. Reprinted with permission from Ref. 35, *Electrochimica Acta* **33,** P. G. Bruce and F. Krok, Copyright 1988, Pergamon Press Plc.

electrode swelled during discharge. This effect may explain the decrease in capacity on initial cycling for cells using V_6O_{13} as a cathode material.[10] The FeS electrode was found to swell considerably on discharge and the particles completely broke up, particularly toward the end of the discharge.

Owens and co-workers[39,40] have studied the cell $Li|P(EO)_8LiCF_3SO_3|V_6O_{13}$ with respect to corrosion phenomena at the interfaces during overcharge. Under

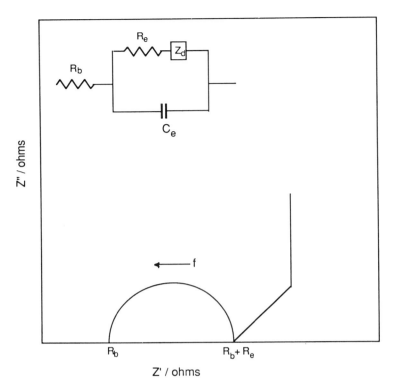

Figure 10.12. Equivalent circuit and complex impedance plot for polymer electrolyte–intercalation electrode interface.

normal conditions, the cell components show no evidence of corrosion; however, severe degradation was shown by both optical and scanning electron microscopy on overcharge. For cells using thick (50–75 μm) cathodes,[40] this was detrimental as the cell voltage reached about 5 V, sufficient to degrade the polymer electrolyte. Deposition of salt was found to occur at the cathode–electrolyte interface. For thin (10–15 μm) cathode cells,[39] it was noted that in addition to this migration, the aluminum backing for the cathode corroded and migrated to the Li surface. It was concluded that polymer electrolyte cells can only withstand approximately 15 to 20% of overcharge.

As the twin channels in V_6O_{13} are surrounded by close-packed oxygen, the mobility of Li^+ ions is expected to be restricted to the (0,1,0) direction. In a study of V_6O_{13} single crystals,[41] the diffusion coefficients at 120°C and $x = 0.17$ have been determined to be 3.5×10^{-8} and 4.9×10^{-9} cm^2 s^{-1} perpendicular to the (0,1,0) and (0,0,1) crystal faces, respectively. Values of 2×10^{-8} and 2.5 cm^2 s^{-1} have been reported for $x = 0.7$ and 2.5, respectively, at ambient temperatures. Munshi et al.[42] have shown that zinc and copper, as well as lithium, may be inserted and removed reversibly from the V_6O_{13} single-crystal cathode.

To investigate the effects of specific lattice orientations on electrode–electrolyte interfacial properties, single-crystal V_6O_{13}–electrolyte interfaces have been studied by ac impedance spectroscopy.[37,41] Jacobsen et al.[41] used 1 mol dm^{-3} propylene carbonate solutions of $LiClO_4$ and $LiAsF_6$ as electrolyte and investigated the interface with the $(0,1,0)$ and $(0,0,1)$ surfaces perpendicular and parallel to the cavity channels, and the surface perpendicular to the a axis also parallel to the cavity channels. Results indicated that Li^+ diffusion in $Li_xV_6O_{13}$ is one-dimensional, proceeding exclusively along the cavity channels in the b axis direction, with diffusion coefficients given as 6×10^{-8} to 4×10^{-9} cm^2 s^{-1}. Bruce et al.[37] have studied the low-molecular-weight $P(EO)_{20}LiCF_3SO_3$ system in contact with single-crystal V_6O_{13} oriented again so that the intercalation channels lay parallel or perpendicular to the electrolyte surface. Figure 10.13 exhibits behavior associated with rough blocking electrodes: an inclined spike at high frequency increasing in steepness at lower frequency. This implies that lithium intercalation does not occur perpendicular to the $(0,1,0)$ crystallographic plane. Where the diffusion channels are parallel to the field direction in the electrolyte (Figure 10.14) a high-frequency semicircle is observed. Direct-current polarization studies confirmed that Li^+ ions may be inserted and removed from the electrode. No evidence for a passivating layer was detected with this purified polymer.

Another aspect of intercalation chemistry is the use of a redox polymer like polyviologen [poly(decaviologen) $(DV^{2+}),2X^-$],

$$-[-(CH_2)_{10}- \overset{+}{N} -\langle\bigcirc\rangle-\langle\bigcirc\rangle- \overset{+}{N} -]-$$
$$X^- \qquad\qquad\qquad\qquad X^-$$

poly(decaviologen)

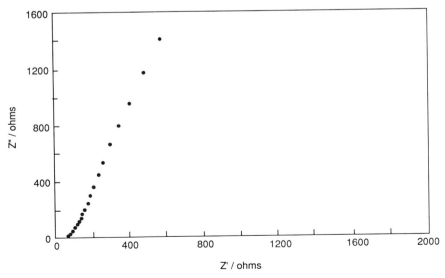

Figure 10.13. $P(EO)_{20}LiClO_4/V_6O_{13}$ single-crystal interface at room temperature. Diffusion channels perpendicular to current flow.

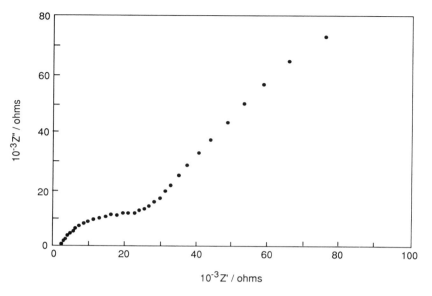

Figure 10.14. $P(EO)_{20}LiClO_4/V_6O_{13}$ single-crystal interface at room temperature. Diffusion channels parallel to current flow.

which has been studied in an all-solid-state cell $Li|P(EO)_nMX|$polyviologen with various MX and n.[43] The electroactivity of the viologen is reversible with formation of the radical cation and the neutral species:

$$(DV^{2+},2X^-) + e^-_{(electrode)} \rightleftharpoons (DV^{\cdot+},X^-) + X^- + e^-_{(electrode)} \rightleftharpoons$$
$$(DV^0) + X^-$$

The polyviologen electrode can thus be used as an anion-specific electrode and has been used successfully to determine some ionic transport parameters.

10.4 Electrochromism of Intercalation Compounds

Electrochromism can be defined as a persistent but reversible optical change produced electrochemically. Of the cathodic coloring materials (i.e., color by a reduction process), WO_3 has received by far the most attention and likewise iridium oxide of the anodic electrochromic materials. The structures of the tungsten trioxides and tungsten bronzes are built from WO_6 octahedra that share corners. These can be arranged in a variety of ways, the more common of which are shown in Figure 10.15.[44,45] Cubic M_xWO_3 structures are formed by the smallest cations only, that is, $M = $ Li, Na, H. The tetragonal and hexagonal structures have larger tunnels and can therefore accommodate larger ions; however, the number of vacant sites per tungsten is reduced. Figure 10.16 shows elements that are known to intercalate into the WO_3 structure. WO_3 dissolves rapidly in aqueous electrolytes and therefore suitable

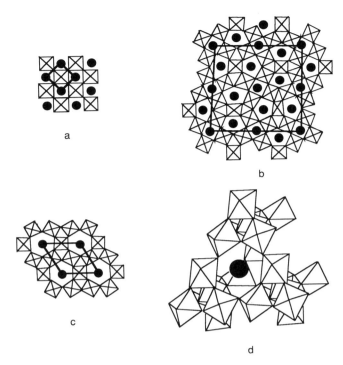

Figure 10.15. Structures of tungsten bronzes: (**a**) Perovskite, (**b**) tetragonal, (**c**) hexagonal, (**d**) pyrochlore.

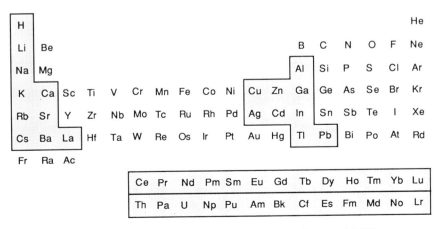

Figure 10.16. Elements known to form tungsten bronzes M_xWO_3.

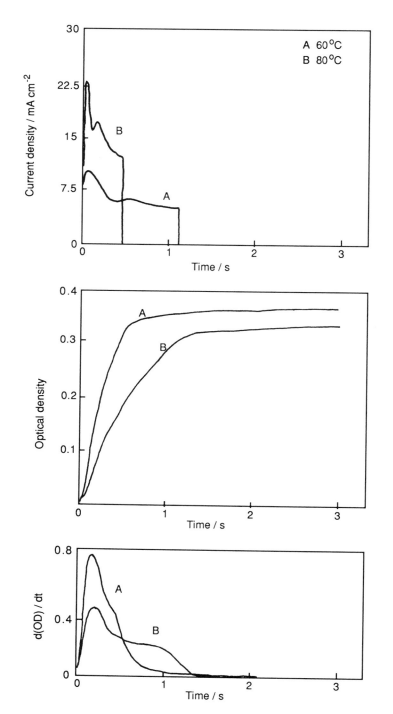

Figure 10.17. Current density, OD variation, and rate of coloration versus time of WO_3 electrode at constant potential (2 V versus Li) at different temperatures. Electrolyte is $P(EO)_8LiClO_4$. From Ref. 50.

organic solvents such as propylene carbonate are often employed. It is important to exclude water to prevent corrosion and deterioration of the cell. Yoshiiki et al.[46] showed by cyclic voltammetry that degradation of WO_3 was accelerated by $LiClO_4$ in propylene carbonate with trace amounts of water. A negative shift of the cathodic reaction potential of the film was explained by a mechanism where H_2O reacts first to form $W(OH)_x$ and then Li^+ exchanges with H^+ of $W(OH)_x$ and results in the formation of Li_2WO_4.

WO_3 has been shown to be chemically and electrochemically stable in contact with a polyvinylpyrrolidone (PVP)–H_3PO_4 electrolyte and a PEO–H_3PO_4 electrolyte[47,48] over a period of 1 year. When a point on the display is colored, however, lateral diffusion of protons and electrons and/or local formation of a concentration cell occurs. This effect was very slow, noticeable only after approximately 3 months. The persistence of open-circuit memory was evaluated by Kobayashi[49] for a $WO_3|P(MEO)_7$–$LiClO_4$ cell (MEO = oligo-oxyethylene methacrylate). The relative optical density fell initially to 0.88 of the original value after 3 h but thereafter remained constant. This was attributed to slow diffusion of Li^+ from the WO_3 into the polymer electrolyte to compensate the excess negative charge at the interface. Coupled cyclic voltammetry–optical measurements on a PEO–H_3PO_4 system[47] showed that above -250 mV (versus NHE), proton intercalation is the only reaction. Below this, a side reaction occurred that was concluded to be the formation of H_2 gas. This gas remains localized at the electrode and can reoxidize during a subsequent anodic cycle. Variations in optical density after current interruption show that H_2 itself can become intercalated. Bohnke et al.[50] performed similar experiments using PEO–$LiClO_4$ as electrolyte. They reported that during anodic polarization, two electrochemical reactions occur. This behavior was not observed for a liquid electrolyte and may be the result of some interfacial reaction. The current response with time associated with coloration presents a minimum, then a maximum, and a similar trend is found in the rate of coloration (Figure 10.17). Both protonic and lithium ion-conducting electrolytes show this phenomenon. For the former, the behavior was explained as the formation of molecular hydrogen by diffusion of adsorbed hydrogen; however, this would not affect the rate of coloration, and in the latter case molecular hydrogen is unlikely to be formed. In this instance, the effect was explained by slow kinetics of ion transfer at the interface.

References

1. D. Fauteux, *Solid State Ionics* **17** (1985), 133.

2. R. D. Armstrong and W. I. Archer, *Electrochim. Acta* **26** (1981), 167.

3. B. C. H. Steele, G. E. Lagos, P. C. Spurgens, C. Forsyth, and A. D. Foord, *Solid State Ionics* **9/10** (1983), 391.

4. S. Atlung, K. West, and T. Jacobsen, *J. Electrochem. Soc.* **126** (1979), 1311.

5. G. E. Lagos and B. C. H. Steele, *166th Electrochemical Society Meeting, October 1984, New Orleans*, Abstract 177.

6. M. Hiratani, *Solid State Ionics* **28–30** (1988), 1431.

7. C. A. Vincent, *Prog. Solid State Chem.* **17** (1987), 145.

8. D. Fauteux, J. Prud'homme, and P. E. Harvey, *Solid State Ionics* **28–30** (1988), 923.

9. K. West, B. Zachau-Christiansen, T. Jacobsen, E. Hiott-Lorenzen, and S. Skaarup, *Br. Polym. J.* **20** (1988), 243.

10. C. A. C. Sequeira and A. Hooper, *Solid State Ionics* **9/10** (1983), 1131.

11. F. Bonino, B. Scrosati, A. Selvaggi, J. Evans, and C. A. Vincent, *J. Power Sources* **18** (1986), 75.

12. F. Bonino, B. Scrosati, and A. Selvaggi, *Solid State Ionics* **18/19** (1986), 1050.

13. J. R. Owen, W. C. Maskell, B. C. H. Steele, T. S. Nielsen, and O. T. Sørensen, *Solid State Ionics* **13** (1984), 329.

14. J. Evans and C. A. Vincent, *Chim. Ind.* **69** (1987), 63.

15. A. D. Holding, D. Pletcher, and R. V. H. Jones, *Electrochim. Acta* **34** (1989), 1529.

16. D. Deroo, A. Bouridah, F. Dalard, H. Cheradame, and J. F. LeNest, *Solid State Ionics* **15** (1985), 233.

17. P. Rigaud, Ph.D. thesis, Université Scientifique et Medicale, et Institut Nationale Polytechnique, Grenoble (1980).

18. M. B. Armand, M. J. Duclot, and P. Rigaud, *Solid State Ionics* **3/4** (1981), 429.

19. M. B. Armand, *Proceedings of Electrochemical Society Workshop on Lithium Nonaqueous Batteries,* Cleveland, Ohio (1980).

20. C. A. C. Sequeira, J. M. North, and A. Hooper, *Solid State Ionics* **13** (1984), 175.

21. T. M. A. Abrantes, L. J. Alcacer, and C. A. C. Sequeira, *Solid State Ionics* **18/19** (1986), 315.

22. M. B. Armand, in *Polymer Electrolyte Reviews—1* (J. R. MacCallum and C. A. Vincent, Eds.), Elsevier Applied Science, London (1987), p. 1.

23. M. Armand, *Faraday Discuss. Chem. Soc.* **88** (1989), 65.

24. M. S. Whittingham, *Prog. Solid State Chem.* **12** (1972), 41.

25. D. W. Murphy, F. J. DiSalvo, J. N. Carides, and J. V. Waszczak, *Mater. Res. Bull.* **13** (1980), 1395.

26. M. S. Whittingham and M. B. Dines, *J. Electrochem. Soc.* **124** (1977), 1388.

27. H. Ikeda, T. Saito, and H. Tamura, *Manganese Dioxide Symp. Proc.* **1** (1975), 384.

28. J. O. Besenhard and R. Schöllhorn, *J. Electrochem. Soc.* **124** (1977), 968.

29. W. R. McKinnon and J. R. Dahn, *J. Electrochem. Soc.* **132** (1985), 364.

30. Y. Charbre, P. Deniard, and R. Yazami, *Solid State Ionics* **28–30** (1988), 1153.

31. B. Scrosati, *Br. Polym. J.* **20** (1988), 219.

32. W. B. Johnson and W. L. Worrell, *Solid State Ionics* **5** (1981), 367.

33. A. Patrick, M. Glasse, R. Latham, and R. Linford, *Solid State Ionics* **18/19** (1986), 1063.

34. G. Pistoia, S. Panero, M. Tocci, R. V. Moshtev, and V. Manev, *Solid State Ionics* **13** (1984), 311.

35. P. G. Bruce and F. Krok, *Electrochim. Acta* **33** (1988), 1669.

36. P. G. Bruce and F. Krok, *Solid State Ionics* **36** (1989), 171.

37. P. G. Bruce, E. McGregor, and C. A. Vincent, in *Second International Symposium on Polymer Electrolytes* (B. Scrosati, Ed.), Elsevier, London (1990), p. 357.

38. P. Baudry, M. Armand, M. Gauthier, and J. Masounave, *Solid State Ionics* **28–30** (1988), 1567.

39. M. Z. A. Munshi, R. Gopaliengar, and B. B. Owens, *Solid State Ionics* **27** (1988), 259.

40. R. Gopaliengar, M. Z. A. Munshi, and B. B. Owens, in *Proceedings, Symposium on Lithium Batteries, Honolulu, 1987,* (A. N. Dey, Ed.), Vol. 88-1, Electrochemical Soc., Pennington, N.J. (1988).

41. T. Jacobsen, K. West, B. Zachau-Christiansen, and S. Atlung, *Electrochim. Acta* **30** (1985), 1205.

42. M. Z. A. Munshi, A. Gilmour, B. B. Owens, and W. H. Smyrl, *Proceedings, 174th Meeting of Electrochemical Society, Chicago* (1988).

43. A. Bouridah, F. Dalard, and M. Armand, *Solid State Ionics* **28–30** (1988), 1193.

44. P. G. Dickens and M. S. Whittingham, *Q. Rev. Chem. Soc.* **22** (1968), 30.

45. D. W. Murphy, J. L. Dye, and S. M. Zahurak, *Inorg. Chem.* **22** (1983), 3679.

46. N. Yoshiiki, Y. Mizono, and S. Kondo, *J. Electrochem. Soc.* **131** (1984), 2634.

47. D. Pedone, M. Armand, and D. Deroo, *Solid State Ionics* **28–30** (1988), 1729.

48. M. Armand, D. Deroo, and D. Pedone, in *Solid State Ionic Devices* (B. V. R. Chowdari and S. Radhakrishna, Eds.), World Science, Singapore (1988), p. 515.

49. N. Kobayashi, Ph.D. thesis, Waseda University, Tokyo (1988).

50. O. Bohnke, C. Bohnke, and S. Amal, *Mater. Sci. Eng.* **B3** (1989), 197.

Index